Life in the Megalopolis

The modern metropolis has been called 'the symbol of our times', and life in it epitomizes, for many, modernity itself. But what to make of inherited ideas of modernity when faced with life in Mexico City and São Paulo, two of the largest metropolises in the world? Is their fractured reality, their brutal social contrasts, and the ever-escalating violence faced by their citizens just an intensification of what Engels described in the first in-depth analysis of an industrial metropolis, nineteenth-century Manchester? Or have post-industrial and neo-globalized economies given rise to new forms of urban existence in the so-called developing world?

Life in the Megalopolis: Mexico City and São Paulo investigates how such questions are explored in cultural productions from these two Latin American megalopolises, the focus being on literature, film, popular music, and visual arts. The choice of these two cities in Mexico and Brazil has to do with their comparable sizes – they are the largest in Latin America (about 20 million inhabitants in each metropolitan area), and figure among the five largest in the world. They are also the most globalized of all Latin American cities. They count on the largest investment of foreign capital (banks, offices of multinational companies, etc.), and their severe social problems have been seen as symptomatic of wider trends in Latin America and the world. This book combines close readings of works with a constant reference to theoretical, anthropological, and social studies of these two cities, and builds on received definitions of the concept megalopolis

Life in the Megalopolis is the first book to combine urban-studies theories (particularly Lefebvre, Harvey, and de Certeau) with Benjaminian cultural analyses, and theoretical discussions with close readings of recent cultural works in various media. It is also the first book to compare Mexico City and São Paulo.

Lúcia Sá is Professor of Brazilian Cultural Studies at the University of Manchester.

'Questioning Cities'
Edited by Gary Bridge, *University of Bristol*, UK and Sophie Watson, *The Open University*, UK

The 'Questioning Cities' series brings together an unusual mix of urban scholars under the title. Rather than taking a broadly economic approach, planning approach or more socio-cultural approach, it aims to include titles from a multi-disciplinary field of those interested in critical urban analysis. The series thus includes authors who draw on contemporary social, urban and critical theory to explore different aspects of the city. It is not therefore a series made up of books which are largely case studies of different cities and predominantly descriptive. It seeks instead to extend current debates, through in most cases, excellent empirical work, and to develop sophisticated understandings of the city from a number of disciplines including geography, sociology, politics, planning, cultural studies, philosophy and literature. The series also aims to be thoroughly international where possible, to be innovative, to surprise, and to challenge received wisdom in urban studies. Overall it will encourage a multi-disciplinary and international dialogue always bearing in mind that simple description or empirical observation which is not located within a broader theoretical framework would not – for this series at least – be enough.

Published:

Global Metropolitan
John Rennie Short

Reason in the City of Difference
Gary Bridge

In the Nature of Cities: Urban political ecology and the politics of urban metabolism
Erik Swyngedouw, Maria Kaika, Nik Heynen

Ordinary Cities: Between modernity and development
Jennifer Robinson

Urban Space and Cityscapes
Christoph Lindner

City Publics: The (dis)enchantments of urban encounters
Sophie Watson

Small Cities: Urban experience beyond the metropolis
David Bell and Mark Jayne

Cities and Race: America's new black ghetto
David Wilson

Cities in Globalization: Practices, policies and theories
Peter J Taylor, Ben Derudder, Piet Saey and Frank Witlox

Cities, Nationalism, and Democratization
Scott A Bollens

Life in the Megalopolis: Mexico City and São Paulo
Lúcia Sá

Life in the Megalopolis
Mexico City and São Paulo

Lúcia Sá

Routledge
Taylor & Francis Group

LONDON AND NEW YORK

First published 2007 by Routledge
2 Park Square, Milton Park, Abingdon, Oxon OX14 4RN

Simultaneously published in the USA and Canada
by Routledge
270 Madison Avenue, New York, NY 10016

Routledge is an imprint of the Taylor & Francis Group, an informa business

© 2007 Lúcia Sá

Typeset in Times New Roman by Graphicraft Limited, Hong Kong
Printed and bound in Great Britain by TJ International Ltd, Padstow, Cornwall

British Library Cataloguing in Publication Data
A catalogue record for this book is available from the British Library

Library of Congress Cataloging in Publication Data
Sá, Lúcia.
Life in the megalopolis: Mexico City and São Paulo / Lúcia Sá.
p. cm.
Includes bibliographical references and index.
1. Mexico City (Mexico)–Social conditions. 2. São Paulo (Brazil)–Social conditions.
3. Mexico City (Mexico)–Social life and customs. 4. São Paulo (Brazil)–Social life
and customs. 5. Mexico City (Mexico)–Civilization. 6. São Paulo (Brazil)–Civilization.
I. Title.
HN120.M45S2 2007
307.76′4097253–dc22
2007013324

ISBN10: 0-415-39271-3 (hbk)
ISBN10: 0-415-39272-1 (pbk)
ISBN10: 0-203-08753-4 (ebk)

ISBN13: 978-0-415-39271-6 (hbk)
ISBN13: 978-0-415-39272-3 (pbk)
ISBN13: 978-0-203-08753-4 (ebk)

For Gordon (again)

Contents

List of figures   ix
Acknowledgements   x

1. Approaching the monster   1

   A tale of two cities 1
   Structure of the book 9
   Mexico-Tenochtitlan 9
   São Paulo de Piratininga 12
   Affinity, or not 15

PART I.
Terra incognita   **19**

2. In fragments for the millennium   21

   A day in the life 24
   Machine flow 32
   Collision 40

3. Flânerie   56

   Post-apocalyptic 56
   Urbenauta 69

PART II.
Find your place in the neighbourhood   **77**

4. Barrio/bairro   79

   Tepito and who was here first 85
   Marchanta at the tianguis 94

5.   Capão Redondo and the space of rap   107

     Lyrics with reason 115
     Work in progress, and on line 132

6.   Writing on the wall and other interventions: epilogue
     in a small gallery   145

     Ruins 147
     More ruins 148
     The dancer: a graffiti mural in Mexico City 150
     Front-page news from a Tenochtitlan rooftop 152
     Lomas belles at home 154

     Notes   157
     References   161
     Index   168

List of figures

6.1 Laura Vinci, Exhibition Arte/Cidade III (1997)   148
6.2 Fátima Roque, Cambuci graffiti   150
6.3 The dancer graffiti   151
6.4 'El Ahuizotl', *La jornada* (Mexico) March 4, 1992   153
6.5 Untitled. Daniela Rossell. *Ricas y Famosas*. Hatje Cantz (2003)   155
6.6 Untitled. Daniela Rossell. *Ricas y Famosas*. Hatje Cantz (2003)   155
6.7 Untitled. Daniela Rossell. *Ricas y Famosas*. Hatje Cantz (2003)   156

Acknowledgements

In order to recover his *muriaquitã*, the eponymous Mário de Andrade travelled from one end of Brazil to the other, the Amazon to São Paulo, and in a sense, having completed *Rain Forest Literatures* (2004), this is what I have attempted to do in this study. The main difference among the obvious many is that, São Paulo being my native city in the first place, I had become fascinated by Mexico City and was given much help along the way. Thanks for that help go institutionally to the universities where I have worked (Stanford) and now work (Manchester). Honouring me with a fellowship in 2003, the Stanford Urban Studies Program gave me a very generous grant which enabled me to travel more than once to São Paulo and Mexico City and buy large numbers of books, films and other materials, as recent as they were indispensable. In Manchester, I am very grateful to the Department of Spanish, Portuguese, and Latin American Studies for allowing me to have a leave at the beginning of my appointment, and to the understanding of colleagues who saw me anxious to meet the deadline set for the first draft and adapted their own schedules accordingly. Chief among them are: Chris Perriam, John Gledhill, Hilary Owen, Sasha Schell, Nuria Triana Toribio, Fernanda Peñalosa, and João Cesar de Castro Rocha.

In Mexico City, I was immensely guided and helped by Margo Glantz (generous and consummate hostess) and many others no less generous with their time, advice and living space, Margarita Orellana and Alberto Ruy, Cuauhtemoc Medina; Nestor García Canclini who was kind enough to let me browse his superb library at length, Johannes Neurat, and Paulina Alcocer.

As the good friends they are, Mary Pratt and Roland Greene read early drafts with keen eyes and Mary did more than a fair share of counselling. It is hard to describe the debt I owe to conversations with colleagues in the San Francisco Bay area, Julio Ramos, Richard Rosa, Alicia Rios, Yvonne Yarbro-Bejarano, Gonzalo Aguilar, and María Helena Rueda. I am grateful for valuable hints and pointers kindly given at conferences and talks, by Marisa Lajolo, John Kraniauskas, and Rossana Reguillo. At the Cambridge conference on The City in 2005 (at which he very helpfully found me a last-minute place), and during his year at Stanford, Paul Julian Smith has been the enviably learned source of ideas that his books declare him to be.

I am particularly grateful to Fátima Roque (Faró) for the beautiful photograph she made especially for the cover of this book: finally, after thirty years, we are able to collaborate! She also took me to special places in São Paulo and gave me many photographs of graffiti, one of which is reproduced in the Epilogue. Antonio Iannarone's photographs of Mexico City graffiti were also a source of inspiration.

I thank Tim Butler for encouraging me to get in touch with Routledge in the first place. My sister Cândida and my brother Antonio have unstintingly helped the project along in many practical ways. And without the help and support of my husband Gordon Brotherston, this book, which I dedicate to him, would simply not exist.

1 Approaching the monster

E foste um difícil começo
Afasto o que não conheço
e quem vem de outro sonho feliz de cidade
aprende depressa a chamar-te de realidade
porque és o avesso do avesso do avesso do avesso
(Caetano Veloso, 'Sampa')

¿Hasta cuando, en qué islote sin presagios,
hallaremos la paz para las aguas
tan sangrientas tan sucias tan remotas,
tan subterraneamente ya virtuales
de nuestro pobre lago y cenagoso
ojo de volcanes, dios del valle
que nadie vio de frente y cuyo nombre
los antiguos callaron?
(José Emilio Pacheco, *El reposo del Fuego*)[1]

A TALE OF TWO CITIES

I had been living in the United States for five years when I first went to Mexico
City, in 1995. For a Brazilian from São Paulo, walking in the streets of the
great capital provoked a certain sensation of *déjà vu*: I felt I was somehow
back home. Sidewalks lined with vendors and streets filled with cars that hardly
moved made me realize that both cities had the same strange rhythm.
Modern wide avenues sided with concrete pointed to what seemed to be a
cult of speed and movement. Yet the actual pace of both cities most hours
of the day appeared to be a cinematic slow motion. Surely, traffic jams are
not exclusive to Mexico or São Paulo, but there was something in the scale
of these two cities that made me feel they had a special place there. Not many
cities in the west have been cut across by such a number of wide avenues
which consistently do not manage to resolve traffic problems. Fewer still combine the heavy traffic in wide avenues with the incessant weaving of thousands of pedestrians through streets crammed with a multitude of stalls and

blankets laid out to display the latest consumer dreams. Then there are the loud voices common to most Latin American culture, a certain sense of humour and, above all, the striking poverty of street children and peripheral slums combined with the glossy modernity of tall corporate buildings. From all those points of view, these two specimens of Latin American megalopolis might as well have been the same city.

It was only when I arrived at the *Centro Histórico* (the old centre) that I felt how different the two megalopolis can be. At the corner of the Zócalo (the enormous open space in the heart of the city) there stood the ruins of an Aztec temple, the archaeological site named Templo Mayor. The half-buried remains of a pyramid revealed several layers of construction: every fifty-two years the Aztecs enclosed the old temple with a new shell. The presence of time was overpowering: I could not ignore that I was in an ancient urban centre, a city erected over what had been, at the time of European arrival, probably the largest city in the world – Tenochtitlan. Having gone down the Paseo de la Reforma south-west to the magnificent Museo Nacional de Antropología, I later saw groups of schoolchildren being told about pre-Columbian gods and rituals. Some of them looked at the statues with admiration, others made jokes, but all of them treated the solemn objects that surrounded them with the familiarity one treats one's own ancestors. As they approached a large display case showing an enormous model of the city's pre-Columbian market, many of them screamed with pleasure, recognizing the fruits, the vegetables, the foods and the artifacts: recognizing, in other words, the presence of Aztec culture in their daily current lives.

São Paulo, by contrast, gives the constant sensation of being a new city. Even its 'foundation building', the tourist attraction Pateo do Colégio, is a mere replica (built in 1954) of the 1554 Jesuit Indian school: the original building was torn down in 1882 in order to make way for the new government headquarters – only one wall of the first construction still remains. Most histories of São Paulo emphasize how successive waves of urbanization have tended to destroy the city's architectural markers and replace them with new ones (Campos 2000: 18). The city, in any case, was an insignificant provincial town until the middle of the nineteenth century, when the combination of large revenues from coffee exports, the centralization of the national shipping (which turned Santos, one hour away from São Paulo, into the busiest port in Brazil), and the replacement of slave labour with large intakes of European immigrants made it grow into an unrecognizable modern city.

This is a tale of two cities – not the London and Paris commemorated by Dickens, but the two largest Latin American cities. Under the name Tenochtitlan, one of them, was the capital of the largest empire in North America when Cortés invaded, and has been a centre of power uninterruptedly for several centuries, now the country's *Distrito Federal* (DF). The other, essentially a twentieth-century city, has never been a centre of official power, although as a matter of fact it has provided the nation with many presidents and has been at its economic forefront for many decades.[2]

Both Mexico City and São Paulo figure among the largest megacities in the world. Rankings of most populous cities vary considerably. United Nations data from 2005 presents Mexico City as the second largest urban centre, and São Paulo as the fourth. In the 2006 data provided by Google Earth, Mexico City also appears as the second largest Metropolitan area, but São Paulo figures as fifth. The German site 'City Population' presents Mexico City as the third largest urban centre, and São Paulo as seventh, while the site 'City Mayors' has São Paulo as third largest and Mexico City as fifth.[3] The variations depend on many factors, the most important of them being whether the rankings use actual census data or projections, and how the 'urban area' is being determined geographically. As Edward Soja points out:

> One of the characteristic features of the megacity is the difficulty of delineating its outer boundaries and hence of accurately estimating its population and size. How many urban centres does one include within the megacity region? How far does the regional hinterland stretch? How far does one go to recognize the increasingly global reach of the megacity?

> (2000: 235)

Whatever their rank among the world's largest cities, both Mexico City and São Paulo are unquestionably a megalopolis (the form conventionally understood to denote that phenomenon in both the singular and the plural), their urban areas each being home to about twenty million inhabitants, and unrivalled in size in Latin America. They also meet other criteria that have been considered essential by some authors for ranking a city as a megalopolis. Jean Gottmann, for instance, in his 1961 classic *Megalopolis*, defined the term not simply as a very large city, but as a region composed of many cities of different sizes that are functionally interrelated – the 'corridor' between Boston and Washington, DC, in the United States, being his case study. Both São Paulo and Mexico City are located in urban areas that include interrelated cities of different sizes.

For Edward Soja,

> [megacity,] the first entry in the glossary of neologisms being used to characterize the spatial transformation of the modern metropolis, refers both to the enormous population size of the world's largest urban agglomerations, and to their increasingly discontinuous, fragmented, polycentric, and almost kaleidoscopic socio-spacial structure.

> (ibid.: 235)

This description is of course eminently applicable to the megacities Mexico City and São Paulo in Latin America. Both are 'global cities', too, in the sense proposed by Saskia Sassen, that is, they concentrate the financial, managerial, coordination and servicing operations of international firms whose economic activities in the last decades have become at one and the same time spatially dispersed and globally integrated. As a result, São Paulo and Mexico City have gone through the economic changes also suffered by other global cities: an increase of inequality based on the simultaneous growth of extremely high-paid executives and very low-paid service workers (and a decrease of jobs in the middle-pay range, whose predominance characterized the post-war industrial economy). These economic and social changes have also had an impact on the space of these two cities, with processes of gentrification and the building of extremely exclusive and heavily secured condominiums (the 'fortified enclaves' described by Teresa Caldeira in *City of Walls*, 2000) on the one hand, and an exponential growth in the number of slums and very poor neighbourhoods on the other.

This book looks at how these processes are represented, discussed, and interfered with, by cultural products from Mexico City and São Paulo – more specifically by works of literature, film, popular music and, and to a lesser degree, visual arts. It concentrates on cultural production from the 1980s to the present, because this is the quarter century that consolidates the processes described above by Sassen: '[In the 1980s] The geography and composition of the global economy changed so as to produce a complex duality: a spatially dispersed, yet globally integrated organization of economic activity' (2001: 3). What resulted from these changes is often referred to as 'postindustrial' or 'post-Fordist' economics. Mollenkopf and Castells alert their readers to the problems imbedded in the term 'postindustrial', but nevertheless continue to find it useful:

> Our use of *postindustrial* implies neither that industrial production has become economically irrelevant nor that control over knowledge has replaced return on investment as the organizing principle of the world capitalist economy. But the term does capture a crucial aspect of how large cities are being transformed: employment has shifted massively away from manufacturing (and handling goods more generally) toward corporate, public, and nonprofit services; occupations similarly shifted from manual workers to managers, professionals, secretaries, and service workers.
>
> (1991: 6)

David Harvey, in turn, refers to the 'ubiquity and volatility of money' caused by the predominance of finance capital in the economy of the 1980s:

> What emerged out of the watershed of the 1970s is not so much an overall flexibility of labour markets, as an unprecedented autonomy of money capital from the circuits of material production – a hypertrophy of finance, which is the other underlying basis of postmodern experience and representation.
>
> (2001: 15)[4]

Indeed, the massive economic and population growth of São Paulo and Mexico City in the 1960s and 1970s was largely related to the status of both cities as providers of industrial jobs. Immigrants from rural regions of the country moved in huge numbers to these cities, settling in poor areas on the periphery (Ciudad Nezahualcoyotl to the east of Centro Histórico in Mexico City, or Capão Redondo, on the southern rim of São Paulo are examples that will appear later in the book), which were transformed into enormous slums. Cultural production from the two cities in the 1960s and 1970s often refers to the urban changes that were taking place in them: massive building, speculation, elimination of green areas, the opening up of wide avenues and the consequent invasion of motor traffic, and the massive influx of immigrants, this time from impoverished rural areas within the nation, exemplarily Brazil's North-east.

Mexican writers from the movement *La onda*, as Jean Franco says, 'tapped into the nervous system of this rapidly changing Mexico city. In Gustavo Sainz's best selling novel *Gazapo*, the city is already losing its familiar contours' (2002: 86). Elena Poniatowska, in a text published in 1980 (at the very end, therefore, of the period in question) celebrates the host of immigrants that were arriving in Mexico City as the 'city's angels' (Ángeles de la ciudad'). *O Homem que Virou Suco* (The man who turned into juice), a Brazilian film of the same year by João Batista de Andrade, depicts the vicissitudes of a North-eastern immigrant in the cruel, cold city of São Paulo.

In the 1980s and 1990s, economic crises and de-industrialization resulted in massive job losses. Former manual workers had to find a means of survival in the growing 'informal economy'.[5] This process is not dissimilar to what was happening in cities throughout the world: automation; the simultaneous spatial dispersion and global integration of large companies, as described by Sassen; and a general shift from production to finance capital. But as Mike Davis points out, in the case of Latin America, Africa, the Middle East, and much of South Asia, the changes had to do no less with the imposition of the IMF economic policies (2006: 11). These processes are described, discussed, and incorporated into cultural products from these two megalopolis since the 1980s.

My use of the expression 'cultural products' is first of all pragmatic: it allows me to refer collectively to the various artistic forms discussed here: literature, music, cinema, and visual arts. Moreover, it has the advantage of reminding us that these works are 'products' in the sense that they are the result of certain economic and social conditions, and that they are sellable, as well as commodities. This does not mean that I see art merely as the result of economic and social conditions, or as mere commodity: cultural, aesthetic and moral values are invested in it (as they are, indeed in other commodities), not to mention pleasure.

The questions that have instigated this analysis are all related to what Julio Ramos (2001) (referring to the nineteenth century) called the 'divergent modernities' of Latin America. Many of the works studied here could probably be described as 'postmodern': they are de-centred or multi-centred, fragmentary, suspicious of 'master narratives', and ultimately the result of post-industrial economic conditions that inform what Fredric Jameson, in the title of an early book on the subject (1991), called 'the logic of late capitalism'. I nonetheless consciously avoid the use of the term. The whole debate on post-modernity in the last quarter-century has produced some fine thinking, but it has also created a series of formulas on how to recognize and describe 'postmodern' works that are most unhelpful to the analyses I make here. Above all, as we will see in Gilles Deleuze's references to Joyce and Mallarmé, multi-centredness, fragmentation, and suspicion of 'master narratives' were already at the core of modernism's most radical statements. What interests me is not so much whether cultural productions, or cities, are 'modern' or 'postmodern', but how 'modernity', and the modernist city, are debated, contested, modified, and doubted from within. The modern metropolis has been called 'the symbol of our times',[6] and life in it epitomizes, for many, modernity itself. Yet, what to make of inherited ideas of modernity when faced with life in Mexico City and São Paulo? Is their fractured reality, their brutal social contrasts, and the ever-escalating violence faced by their citizens just an intensification of what Engels (1993) described in the first in-depth analysis of an industrial metropolis, nineteenth-century Manchester? Or have post-industrial and neo-globalized economies given rise to 'spaces of insurgent citizenship' against the modernist city, as Holston suggests (1999: 157)? How do cultural products of these megalopolis adapt (or not) to forms often associated with the representation of modern cities in the late nineteenth and early twentieth century, most notably the fragment? Is the artistic representation of speed (a well-known modernist ambition) still desired by the artists of the megalopolis? If so, how does it relate to the experience of unprecedented congestion, on the one hand and, on the other, unprecedented possibilities of movement, synchronicity, and cybertravel? Similar questions apply to the quintessentially urban figure of the *flâneur*, and how he (or she) is revisited, re-created and denied by recent literature, cinema, and popular music of Mexico City and São Paulo.

These questions refer to the city as a whole, and to the difficulties posed by any attempt to represent, imagine, or even physically to map, the megalopolis. They are matched here in the subsequent set of enquiries, which deals with specific localities in Mexico City and São Paulo, or, more precisely, with the concept of locality itself: how *barrios* or *bairros* (i.e. neighbourhoods) are represented and constructed in the cultural products of these two cities. Or, conversely, what role do artistic forms play in the production of neighbourhoods?

Needless to say, this book does not presume to provide answers to all of these questions. Rather, by looking at specific works, it examines how each of them deals with particular questions – how, in other words, the works themselves re-cast the questions and attempt, some times, to provide answers, while at others resolutely refusing to do so. By looking closely at carefully selected works of literature, songs, and films from these two megalopolis, we can understand some of the ways in which 'monster cities' are experienced as everyday life, how they affect their inhabitants and how reciprocally they are affected by them. It is to be hoped that these cultural products will help us better understand some of the processes involved in the 'life in the megalopolis', besides whatever pleasure they may give or interest they may arouse. To quote Paul Klee (cited in turn by Henri Lefebvre): 'art does not reflect the visible: it renders visible' (1991: 123). Conversely, the understanding of urban theories and concepts from the field of urban studies may help illuminate, as well, some of Mexico City's and São Paulo's best, or at least most significant, contemporary cultural production.

Crucial to this study is the concept of *space*, understood here, after Henri Lefebvre, as 'social space', that is, a 'social product' (ibid.: 26). Megalopolis like Mexico City and São Paulo are the result of complex human relationships – socio-economic, political, cultural, educational, artistic, and so forth. They also influence and impinge on human relationships. As Lefebvre says, 'space produces and is produced' (ibid.: 85). As well as Lefebvre's, David Harvey's space-centred re-readings of Marxist theory gave me a better understanding of cities – particularly of the ways in which urban space reproduces and feeds social inequality. Michel de Certeau's (1984) explanations of how everyday practices like walking and consuming produce meaning are likewise important for the analyses I make in this book, not so much because of any attempts to establish a 'rhetoric' of walking or other spatial practices (I make no such attempts) but because they allowed me to see that it was possible to find meanings in apparently meaningless urban acts. The influence of these (and other) urban theorists is exposed to the lightning-flash fragmentary texts of Walter Benjamin (1999a, 1999b), which guide me through the readings of megalopolitan *flâneurs* and other aspects of life in the two megalopolis.

Recent works on Mexico City and São Paulo by sociologists, anthropologists, and geographers were also of great importance for the conception of this book. This is particularly the case with Nestor García Canclini's (1998) analyses of consumerism and other cultural processes in Mexico City; Rossana Reguillo's (2000) studies of youth groups in the same city; Teresa Pires Caldeira's (2000) seminal *City of Walls*, and Rogério Proença Leite's *Contra-usos da Cidade* (2004), and many other works of this kind, which provided invaluable information and insights. These authors tend, however, to pay little or no attention to culture, and when they do (as is the case with García Canclini and Rossana Reguillo Cruz), it is in numerical terms or through interviews with individuals; no attempts are made to generate analyses of works in any media. This book therefore is complementary to theirs: I have not interviewed inhabitants or artists in either city, and I did not collect any numerical data, precisely because so many other works on these two cities have done so already. My readings of Mexico City and São Paulo are based on close analyses of particular works, relating them to the economic, social, and political context in which they were produced and/or on which they impinge. My aim has been to let the works, as much as possible, speak for themselves.

In order to choose the particular examples I concentrate on in this book, I spent most of my research time over three years collecting, reading, viewing, and listening to an immense number of works in several media. Dealing with large amounts of very recent works was often daunting – for the most part I could not base myself on previous critics, and had no guide to indicate the quality or, most important, the relevance of particular pieces, especially in the case of literature and popular music. Since in Mexico political and cultural life is much more centralized in its capital city, I expected to find much material there, and did. In the case of São Paulo I confess I was surprised by its immense production, especially of literature. For the most part, books were acquired on location, at bookstores in Mexico City and São Paulo, most often in fact at Ghandi in Mexico and at Livraria Cultura in São Paulo. If I name them here it is in order to highlight the importance, for research of this kind, of the quintessentially urban activity of book browsing, and the luxury and pleasure of having in earshot a very well read and informed sales person (threatened as this style of professional may have become most recently by the further inroads of managerial capitalism). Compact discs of popular music were also mostly bought on location. In the case of cinema and visual arts, the choice of works was evidently much smaller and I was generally able to learn about them before seeing them, through reviews in newspapers or magazines, and from what people I trusted who knew that scene were saying. The final list of works I analyse here was determined by the themes I wanted to discuss in the book. To take salient examples: those works that represent the most original attempts to make use of fragmentary form to characterize the giant city; or which illustrate particularly megalopolitan varieties of *flânerie*.

For Walter Benjamin, we could divide 'all the existing descriptions of cities into two groups according to the birthplace of the authors': those written by natives, and those written by outsiders, and the ones written by outsiders, he reminds us, are far greater in quantity:

> The superficial pretext – the exotic and the picturesque – appeals only to the outsider. To depict a city as a native would call for other, deeper motives – the motives of the person who journeys into the past, rather than to foreign parts. The account of a city given by a native will always have something in common with memories; it is no accident that the writer has spent his childhood there.
>
> (1999b: 262)

This book is both those things, and a homage in both cases. It is a recollection and an attempt to re-collect my native city, São Paulo, constantly and purposefully cross-referenced with reactions to a city that I know only as an outsider, Mexico City. Although it is based on works by authors from and in those places, my interpretations of the works themselves are necessarily guided by my own experiences of the cities in question. In the case of

São Paulo, the city where I lived until I was 30 years old, and to which I return every year to find smells, light, sounds and 'infinite shades of gray' (to quote a line by Bruno Zeni (2002), probably inspired by de Chirico) that are still more familiar to me than the sounds, smells, light or colours of any other place. In the case of Mexico, a city that I can see only with the eyes of a *paulista*: the eyes of someone who is permanently fascinated by the similarities and mesmerized by the differences between it and my native city.

STRUCTURE OF THE BOOK

The remaining pages of this introductory chapter offer brief historical accounts of how each of the two cities grew to be what it is in its own way, and key similarities and differences between them. The main body of the book falls into two subsequent parts, each having two chapters. Part I, 'Terra incognita', looks at representations of the city as whole, and opens by centring on the megalopolitan recasting of two modernist tropes: the fragment, and the *flâneur*, which are then explored in depth. Chapter 2, 'In fragments at the millennium' offers a close analysis of attempts to represent the 'unrepresentable' city of the year 2000, in the books *Eles Eram Muitos Cavalos* by Luiz Ruffato and *O Fluxo Silencioso das Máquinas* by Bruno Zeni, and in the blockbuster film *Amores perros* directed by Alejandro González Iñárritu. Chapter 3, 'Flânerie', focuses on that Baudelairean figure as he/ she reappears in the megalopolis in various and strange forms. The works in question are *crónicas* by the Mexicans Armando Ramírez and Carlos Monsiváis, and the weekly television programme *Aquí nos tocó vivir*, by Cristina Pacheco. In the case of São Paulo, they are the short film *Opressão*, by Mirella Martinelli, and the travelogue by the *urbenauta* Eduardo Emílio Fenianos, *São Paulo: uma Aventura Radical*.

Part II, 'Find your place in the neighbourhood', concentrates on specific places within the megalopolis and on the efforts that local inhabitants have made to represent and interfere in its life. Chapter 4, '*Barrio/bairro*', discusses how Armando Ramírez geographically inscribes Tepito, a *barrio* in the heart of Mexico City, and champions it against the wealthier parts of the city in his novels and *crónicas*. Chapter 5, 'Capão Redondo', examines how rappers, the novelist Ferréz, and contributors to a neighbourhood website are producing their own space on the periphery of São Paulo. All this is rounded off with readings of visual art in the final chapter, in the form of an epilogue, 'Writing on the wall and other interventions'.

MEXICO-TENOCHTITLAN

According to their own records, the Aztec 'heron people' who came to prefer to be called Mexica, founded their capital Tenochtitlan, today Mexico City, on a lake island in the year 1325 (2 House in their calendar). This accomplished, they quickly launched themselves on the road to empire, initially patronized by the Culhua whose name Columbus heard far away in the Caribbean (their town Colhuacan remains readily locatable to the south of the centre). This we learn from the codices or books of Mexico and Mesoamerica, written in the script known in their language, Nahuatl, as *tlacuilolli*. The opening page of the Mendoza Codex situates the epochal foundation in the model of quarters. In practice, this model determined the way the city was laid out (later called its 'traza' in Spanish) and, as the empire grew, the boundaries were established between the four provinces that surrounded the metropolitan centre. Between what were seen as the seas to south and north, Pacific and Atlantic ('mar del sur, mar del norte' in Spanish), the empire spread Nahuatl wide as its lingua franca, especially after the taking in 1467 of Coixtlahuaca, which controlled the tribute highway to the east, to what remains the frontier with Guatemala (the Nahuatl name for that Maya territory). This is also true of the Mexica version of *tlacuilolli* script which had among its functions the recording of tribute demanded and received, though as a script as such *tlacuilolli* can be traced back over many centuries to beginnings among the Olmec and Maya.[7]

On August 13, 1521, the Spaniards, led by Hernán Cortés, finally forced Tenochtitlan to surrender after a long and brutal siege, and set about destroying palaces and temples not already reduced to rubble by their cannons. A few months later, Cortés and his men had to decide where to locate the capital of the newly conquered domain. They discussed whether to place it in the areas of Coyoacan or Tacubaya or further afield, but for security reasons (and presumably to be able to use the urban structure already available to them), they chose to build it on the site of Tenochtitlan, keeping its original name. The architect responsible for the project of reconstruction was Alonso García Bravo (Espinosa López 1991: 4; Mier y Terán Rocha 2005: 106). In the words of Mier y Terán Rocha:

> The Spaniards presented themselves with a project to conquest the Aztec Empire which consisted in appropriating the space of its city, its capital, its symbol, turning it into the focus of colonial settling . . . It meant imposing a new culture that would express itself in the new colonial city of Mexico, which was born from the ruins of what had been a great metropolis, and keeping its urban conditioning.
>
> (2005: 101; my translation)

Before very long, it was the capital of what was envisaged and became the vice-royalty of Nueva España.

In the first decades, the Spaniards kept to the original *traza* of the city, but made considerable changes to its environment, above all in terms of existing dikes and canals built to control the lake waters, and the European innovations have been seen as responsible for the frequent floods that began to plague the city (the first great flood happened in 1555). As Espinosa López points out, the floods were aggravated by general alterations to the landscape and ecological changes implemented by the Spaniards themselves. These brought, above all, the intensification of dry agriculture, as opposed to the sophisticated lakeside *chinampa* system that had worked so well for the Aztecs and their predecessors and which can still be witnessed in Xochimilco, a main agriculture supplier for the city up to the nineteenth century. The other innovation was the large-scale introduction of domesticated herd animals unknown in America, cattle-capital which heartily chewed and trampled down unfenced local fields:

> Bernal Díaz reminded us that in forty years of Spanish government, the fauna and flora disappeared without trace; moreover they were growing food on banks where silt displaced water. At the time of the Indians there were no horses, nor cattle, nor did they plough the land and the slopes and hills were not intensely planted.
>
> (Espinosa López 1991: 4)

The centuries of colonial rule were marked both by intense construction and attempts to drain the lake waters. Many palaces, both baroque and neo-classical, were built in the centre of the capital, along with numerous convents and churches of various orders. In the late eighteenth century, Humboldt called the city 'la ciudad de los palácios'.

Independence in 1821 created the D.F. (Federal District). Forty years later President Juárez declared the secularization of Church property. Meanwhile the city grew slowly, and by the end of the nineteenth century, its population was slightly over 500,000. The early twentieth century was marked by the Revolution that toppled the dictator Porfírio Diaz in 1910 and lasted until the promulgation of the new constitution in 1917. From an urban studies point of view, the Revolution can be described as a struggle 'within revolutionary leadership over ideological direction the Revolution was to take and the power to be granted to rural versus urban classes in constructing that path' (Davis 1994: 20). It left a semi-destroyed city, whose faulty infra-structure became the reason for numerous struggles and riots during the 1920s.[8] It also gave rise to a national archaeology, on a scale and with a purpose unmatched in any other American state. The imaginative and practical impact made by Manuel Gamio's excavations at nearby Teotihuacan (a megalopolis in its day) and the official publication of the results in 1922 can still be strongly felt. The same can be said about the work begun in the next decade by Alfonso Reyes at Monte Alban, Oaxaca.

In the 1930s, under President Cárdenas, a process of industrialization started, and the population began to grow, mostly thanks to immigration from rural parts of the country. In 1942 the Federal Government froze rents in the D.F., which became an important factor for keeping the low-waged and the poor in central areas of the city (Tepito being a main example). In 1950, the urban area of the D.F. had reached 3 million people. Industrialization intensified in the 1960s and 1970s, and so did the construction of avenues, the opening of new bus routes, and the building of the Metro (the first line was inaugurated in 1969) to accommodate a population that was by now growing vertiginously, especially towards the poor north-east.

SÃO PAULO DE PIRATININGA

The foundation of São Paulo is shrouded in polemic. Its official history says it was founded by Jesuit priests in 1554, in an act that effectively imposed the name of Christ's missionary to the Corinthians and others upon land named Piratininga in Tupi-Guarani. For his part, however, the early *cronista* Frei Gaspar claimed that by the time the Jesuits arrived, a village with a Christian name was already in place, not too far away: Santo André da Borda do Campo. Disciple Andrew's name was imposed on to local geography thanks to João Ramalho, a Portuguese adventurer about whom little is known, and today its initial A announces what is the ABC of the city's industrial south (the towns of saints André, Bernardo and Caetano). Ramalho is said to have married a local Tupiquinim woman (or several Tupiniquim women, depending on the version) and this can only have fostered his career as a 'founder' of towns. At all events, he was not deemed acceptable by the Jesuits. They alleged reasons of security, and in setting up their school further north at Piratininga, the Jesuits urged the inhabitants of Ramalho's village to come and join them (Silva 1884: 26).

It was the village of São Paulo that most effectively felt the European presence and their power, though ironically not so much through the Jesuits as through adventurers who became fierce rivals, the *bandeirantes*. These people took their name, honestly enough, not from a Christian saint but from the flag or *bandeira* that rallied their expeditions to the interior. During what was left of the sixteenth century and until the early eighteenth century, the *bandeirantes* would set off from São Paulo, well armed for their time, in search of native populations to bring back as slaves, as well as precious metals and precious stones, and it was their drive west towards Paraguay that brought them into direct conflict with the missions set up by the Jesuits. Their leaders, the *bandeirantes*, are still praised as heroes in official history, and serve as a touchstone for *paulista* identity. They were invoked as the spirit of modernity in the Semana de Arte Moderna in 1922, and have become the name of media channels and a university.

São Paulo village was upgraded to a city in 1711, but remained quite small until the mid-1900s. With the arrival of large contingents of European immigrants, especially Italians, the population of São Paulo jumped from 30,000 in 1870 to 240,000 in 1900. This growth was accompanied by a large increase in the territory claimed and adopted by the city, thanks in part to lots (*loteamentos*) reserved for the elites, and to the dumping of enormous numbers of poor people in run-down housing near the centre (the *cortiços*), or on the outskirts (the early *periferia*). Campos observes: 'For each neighbourhood created for the coffee elites, like Campos Elíseos, Higienópolis, and

Paulista Avenue, dozens of areas would be taken over by the poor, and to these we have to add the rather more precariously occupied areas' (2000: 99). The coffee elites made it their priority to urbanize the centre of the city and the 'noble neighbourhoods', in what Campos calls a hegemonic process whose objective was 'to establish urban centrality as a dominant element in the political, social, and economic panoramas' of the region (ibid.: 99–100).

By the late 1870s, and particularly after the proclamation of the Republic in 1889 (a year after slavery was abolished), the city was undergoing major territorial re-definition: 'The emergence of segregation as a structural element in the city was one of the main changes in that period' (Rolnik 1997: 28). The creation of wealthy neighbourhoods was accompanied by laws demanding that houses built in those areas had to be set back from the street and surrounded by gardens. As Rolnik puts it: 'These laws, which defined the ways of building in the elite neighbourhood, are characteristic of creation of urban legislation in the city of São Paulo: the law exists to guarantee the space of the elites' (ibid.: 46). The rural zones around the city and the peripheral areas where many of the poor were moving to remained mostly unlegislated. This trend continued with the institution of new urban laws in 1910, which left much of the periphery unregulated (Caldeira 2000: 216).

Through the alliances they forged with the milk producers of Minas Gerais, the São Paulo elites were able to keep political control not just of the city but of the country for several decades. In the city itself, coffee money, growing industrialization, and groups of ambitious speculators helped to attract foreign investors who, along with local property owners, formed in 1912 the Cia. City, which continued the trend of opening lots exclusive to the elites. In these neighbourhoods, and in the many reforms proposed for the centre in the first decades of the century, São Paulo projected the image of a modern city. By 1920, its industrial park was superior to Rio de Janeiro's, then still the country's capital (Somekh 1997: 70).

It was in this context that in 1922 a group of young men and women, many of them sons and daughters of the coffee and industrial elites, organized the famous Semana de Arte Moderna (Modern Art Week) in the sumptuous Municipal Theatre in the centre in the city (which had been inaugurated not that long before, in 1911). It included readings of avant-garde poetry, exhibitions of paintings and concerts of modern music. One of the organizers, Mário de Andrade, published a collection of poetry in the same year in celebration of São Paulo, entitled *Paulicéia Desvairada* (translated into English as Hallucinated City). The week caused a certain scandal in Brazilian press, and the artists involved in the movement became known as 'modernos paulistas', or 'modernistas' – adjectives that further helped to associate the city with an idea of modernity. In addition, most Brazilian literary histories have hailed the Semana de Arte Moderna as the official landmark, as it were, of avant-garde literature and art in Brazil, although, as many critics have recently pointed out, avant-garde poetry and paintings were being produced before the Semana in other parts of Brazil.

The *modernista* group went on to publish seminal works about the city. In 1928, Mário de Andrade released the novel *Macunaíma*, based on popular narratives and songs, above all native Pemon stories. In it the Amazonian trickster Macunaima travels to São Paulo where he has many adventures and sends a letter back home describing the frantic rhythm of the metropolis where money could buy everything. Oswald de Andrade, the most boisterous member of the Semana, also wrote avant-garde novels and poetry describing the city through the empty lives of its well-heeled bourgeoisie. In contrast, Antonio de Alcântara Machado, son of an elite *paulista* family and closely connected with several participants of the Semana, wrote sympathetic short stories about poor Italian immigrants in the São Paulo of the 1920s. In the same vein, Patrícia Galvão published in 1933 the 'proletarian novel' *Parque Industrial* (Industrial Park) describing the lives of female factory workers in the city.[8] Not part of the *modernista* groups, in 1929, the Hungarian immigrants Adalberto Kemeny and Rudolf Rex Lustig filmed *São Paulo: a Sinfonia da Metrópole*, clearly based on Walter Ruttmann's *Sinfonie einer Großstadt* (1927). Like its German antecedent, Kemeny and Lustig's film shows one day in the life of São Paulo, emphasizing 'its metropolitan character and dynamic, its industrial vocation, and its intense rhythm' (Campos 2000: 461). In 1937, the 30-storey-high Edifício Martinelli was inaugurated, to become a landmark in the verticalization of the city.

Between 1910 and 1930, São Paulo went through several urban renovations, and was endowed with its first modern avenues that updated the nineteenth-century splendour of Avenida Paulista. Then, in 1930, the engineer Francisco Prestes Maia (who would become mayor of the city in 1938) published his 'Avenues Plan' (Plano das Avenidas), an ambitious urbanization project clearly inspired by the Haussmannization of Paris. Through its concentric disposition and its emphasis on the use of buses (which needed less infra-structure than trams and allowed the working classes to live far from work), the plan is mostly responsible for the centre–periphery urban organization that has dominated São Paulo since the 1940s, the decade in which the city reached its first million inhabitants. The model is characterized by relatively low population density (lower, in any case, than that of the first decades of the century thanks to the expansion of the metropolitan area); separation between social classes; and a strong tendency towards home ownership for the poor as well as for the rich (Caldeira 2000: 218). This new urban configuration further aggravated the city's schizophrenic regulatory system: the wealthier central parts of the city continued to be hyper-supervised while the urbanization of *periferia* was left to private owners with no little or no respect for building codes or planning norms (ibid.: 219–23; Maricato 1996).

The celebrations of São Paulo's centenary in 1954 were marked by the slogan 'São Paulo cannot stop' ('São Paulo não pode parar'), a prophecy (or warning) of what in the next two decades would become its vertiginous and chaotic growth (if not default movement along its future traffic lanes). The advent of huge automobile factories (among them Volkswagen, Mercedes,

Ford and General Motors) in ABC and the south-eastern suburbs towards the end of the 1950s brought with it even greater waves of immigrants, this time from the country's northeastern states (who, following the concentric model, settled in the city's *periferias*) and made motor vehicle traffic central to transport. In 1970, a little over eight million people lived in the urban area of São Paulo, a number that had grown to twelve and a half million by 1980 (Brito 2006: 225). The population continued to grow, though the rate at which it did dropped dramatically in the 1980s and 1990s, due to falls in both the birth rate and in immigration from the North-east. The city continues to grow, albeit now at a much slower pace.

AFFINITY, OR NOT

So much for the individual tales, very briefly told here, of Latin America's two megalopolitan cities. Definition of life in them can help associate them as a pair, sharpening the distinctions between them. This exercise depends on multiple factors, not least those highlighted, assumed or resisted in their cultural production over the past quarter century. To begin with, it takes us back deep into geological time.

Horrendous as it is recognized to be in many, if not most, of these recent megalopolitan texts, the heavy air pollution with the attendant thermal inversion that oppresses both cities[9] is typically imagined to be worse because the choice of their location has climatically been a source of pride, high up in the region of 'the most transparent air' at either edge of the American tropics, at more or less equivalent distances from Quito, to the north-west and to the south-east. This air is what encouraged the eagle, who announced to the Aztecs the site of their future capital Mexico-Tenochtitlan, to sun his wings in the fresh early morning, and what gave São Paulo the reputation (in late nineteenth century) of being a healthy, temperate city (in contrast with the hotter Rio de Janeiro, often plagued by tropical malaises). Determined by bedrock, the hydrography in each place has likewise been modified to accommodate human influx, with smelly results that likewise pervade recent literature. The sweet and salt water in the lake basin where Tenochtitlan was founded, expertly managed by imperial engineers before the European invasion, were drained wholesale after it through the north-flowing channel (or cut, *tajo*), soon a huge sewer, which technically converted the basin into the 'Valle de México'. Today it has become the Gran Canal del Desague, ever threatened by noxious back-up due to the sinking of the city that draining the basin has caused. In São Paulo, engineers have succeeded in controlling at will the directional flow of the River Pinheiros, alongside which runs a main ring-road (*marginal*). Yet so far they have not succeed in controlling the river's polluted stench, comparable to that of the Gran Canal, which with no less unpleasantness greets the visitor at the airport (in this case, the Tietê river), a principal node like Mexico DF in global traffic.

Factors that ultimately rest on geological foundations in the American tropics account even more for major differences discerned in today's megalopolis. Mexico City's volcanic rim visibly continues to frame it and cinematic images of it; thick layers of silt from the former lake have allowed the construction of the vast metro network. In both respects granite-based São Paulo differs, with corresponding consequences for such key megalopolitan concepts as periphery and traffic flow. In its day, before and after Cortés invaded and the draining of the lake, the island city of Mexico-Tenotchitlan was criss-crossed with canals (the 'Rios' that are today's traffic arteries) and overwhelmed everyone with its beauty, so that even Venetians compared it favourably to their hometown (at least in their illustrated maps). A major ingredient in the memory of such a city in early colonial times, say, in the early poetry of José Emilio Pacheco (husband of Cristina, and student friend of Carlos Monsiváis), today can add a certain twist to nostalgia for it, and has no counterpart at all in the topography of São Paulo.

It is not only in political terms that Mexico has remained an unquestionable centre of power for many centuries: culturally, too, as Carlos Monsiváis (1995) observed, Mexico is a country of one single city. São Paulo, by contrast, has always had to fight for its importance with Rio de Janeiro, capital of the country until 1960 and up to the 1980s Brazil's most important cultural city. The two metropolis have maintained a complex relationship over the last century, marked by a certain competition and stereotypes. Rio is known by the epithets *cidade maravilhosa* (wonderful city) and 'Brazilian postcard': it is a tourist attraction and the home of a supposedly more 'laid back' population. São Paulo does not claim to be beautiful, nor a tourist destination, but describes itself rather as the 'locomotive' of Brazil: the economic force that leads the country. Its pride has traditionally been its cultural wealth and the energy of what it likes to think as its hard-working, punctual citizens. How these ideas are recast in times of high unemployment will be one of the themes of this book. In any case, the main difference here is that Mexico City has no plausible rival in the national scene.

As my experience of Templo Mayor and the Museo de Nacional Antropología revealed, differences between São Paulo and Mexico City deeper in human time are entirely categorical. To try and rehearse here even the most blatant examples of this difference is out of the question, Mexico-Tenochtitlan's experience being so rich and complex. Evidenced in a host of sources including tlacuilolli manuscripts written before the Europeans invaded, this heritage distinguishes Mexico not just in tropical America but the world as a whole. It constitutes an immensely rich intellectual and cultural bedrock, as readily legible nowhere else in the continent. In terms of its own history, this script or visual language can be connected to millennial antecedents in Cholula and even Mesoamerica's 'mother culture', the Olmec, recognized by the Aztecs as far more ancient and venerable than themselves. (Much nearer to us in time than they were to the Olmec, the Aztecs did archaeological research of their own, excavating Olmec sites and making tourist visits to neighbouring Teotihuacan.[10])

In turn, its force in today's megalopolitan imagination should not be underestimated or quickly dismissed. It is operative in a whole range of works, artistic and literary, and in the graffiti archive. Its most solid ancient configuration, as it were, is the sculpted basalt disk once revered in the Templo Mayor, the massive Piedra de los soles or Sunstone, which intricately inscribed in tlacuilolli superbly images the cosmic scheme of world-ages in which our time is set. The unearthing and recovery of it terminally unsettled the Spanish colony, and after the Revolution it was correspondingly vaunted in the style of mural propaganda which Diego Rivera made famous in the 1920s, being eventually allotted an appropriate place on the high altar of the Museo Nacional de Antropología, in itself a magnificent statement of Mexican (if not Mexica) revolutionary ideology (which it is said allowed no monuments to European invaders to be erected in the capital). It remains to be seen what the stone with its cosmic message might presage now that at the end of the millennium the Temple itself (my first alert) has been excavated.

In day-to-day life in the megalopolis, markers as indispensable as toponyms continue to affirm resident meaning that is far longer standing, and voiced in the tongues that respectively charted domains now being incorporated into the huge economic blocs called TLC or NAFTA in the north and Mercosur to the south. It is audible enough in the Tupi-Guarani that locates main sites in and around São Paulo like Anhangabaú, Tietê, Ibirapuera, and Guarapiranga (and which named neighbouring Paraguay and Uruguay); in Mexico City, Nahuatl can still claim to name the city and the country along with such Central American states as Guatemala and Nicaragua, surrounding states in the federal system (Oaxaca, Chiapas, Nayarit, Colima, etc.), and many of the former towns and districts that now comprise the city, like Azcapotzalco, Tlatelolco, Mixiuhcan, Texcoco, Chapultepec, Colhuacan, Coyoacan, Churubusco, Iztapalapa, and (though small by no means least) Tepito. The great hero of resistance to the European invasion, the Emperor Cuauhtemoc gives his name to the Delegación in which Tepito is found and, perhaps most striking of all, in announcing the stations in the metro network alphabetic transcriptions of these names are matched with their *tlacuilolli* originals, which display memorably for all local eyes, for example, the grasshopper (*chapulin*) of Chapultepec, the coyote of Coyoacan, and the midwife of Mixiuhcan. And it is not only a matter of toponyms: as García Canclini points out, besides having over 200,000 Nahuatl speakers, Mexico City is nowadays perhaps the largest Otomi and Mixtec city, thanks to the immense numbers of Indigenous immigrants from rural areas who have settled there (1998: 20).

I.

Terra incognita

2 In fragments for the millennium

An idea that often comes up in discussions of the megalopolis of today is that of its being beyond apprehension, its inapprehensibility. The city is described as being too vast, and too connected to global phenomena, for any one individual to get to know or even imagine. Edward Soja, partly quoting Ian Chambers, alludes to the difficulties of mapping the megalopolis. 'We can no longer be confident that we know how to map the new metropolis', he says, and 'even such expansive terms as megalopolis and megacity no longer seem sufficient to define the outer limits of the globally restructured metropolitan region'[1] (2000: 218). This is how Nestor García Canclini refers to Mexico City, the place where he lives and works: 'The narratives that organize urban experience in the territory of the historic city disintegrate as the megalopolis becomes inapprehensible.' To strengthen his point with reference to the Mexican capital, by then quite unrecognizable as the place that literature had immortalized as 'la región más transparente del aire', he quotes the writer Juan Villoro:

> In 1928, Carlos Fuentes could still attempt to write a mural-novel that encompassed the city in all its strata: *La región más transparente* (Where the air is clear). Now there would be a need for the combined talents of fifty novelists to recreate the numerous cities that we call Ciudad de México.
> (García Canclini 1998: 23)

In itself, the idea of the inapprehensible city long antedates the modern megalopolis and may plausibly be traced back to the earliest examples of *urbs*. Without question the notion is found fairly frequently in nineteenth-century literature and is well nigh obligatory in representations of the modernist city. Invoking James Joyce, for instance, Raymond Williams observes that in the modernist experience, the city had become internalized: 'In a certain sense there is no longer a city, there is rather a man walking through the city'[2] (1973: 243). And faced with a Paris that was undergoing serious transformation, Baudelaire felt the need to develop 'a new language, a poetic prose, musical beyond rhythm and rhyme, supple yet resistant, capable of adapting to the lyric impulses of the soul, the wave patterns of dreams, the leaps and starts of consciousness' (1965: 54). He stresses, too, that it was chiefly when 'frequenting the largest cities', that is, 'from the mix of the countless experiences' they offer that his obsessive ideal of forging a new language was born (ibid.: 54). Poetry, central to Baudelaire's work, was not alone in being deemed in need of generating fragmentary forms the better to express the experience of the modern city. A devotee of Baudelaire, the theorist Walter Benjamin felt the need to appeal to the fragment form in his *Arcades*, when intuiting how to think the modern city.

As Susan Buck-Morss notes, Benjamin 'believed (differently from Adorno) that the historical-philosophical constellations that he wished to transmit would be better represented by dialectical images than by dialectical argumentation' (1989: 67). Montage, for example, had for Benjamin the power to 'interrupt the context in which it was inserted, and as a result to serve as an antidote to illusion' (ibid.: 67). Fragmentary form in the literature and the cinema of the twentieth century is intrinsically bound to the modern city. As never before, at the beginning of the twentieth century human beings could travel through ever greater distances in ever shorter time. For many modernists, this experience, that of speed, could be represented only through fragmentation. The eyes that observe the city from the train, the tram or the car capture just parts of objects that move by. And the lights and mirrors of the consumer city, reflecting each other, create a phantasmagory (to use the term that Benjamin took from Marx not too long before it entered English) that multiplies the fragments to infinity.

Thanks in part to Walter Benjamin, the figure that came to represent the individual's attempts to apprehend the modern city was Baudelaire's *flâneur*. In Benjamin's words, 'The city is the realization of the ancient dream of humanity, the labyrinth. It is this reality to which the *flâneur*, without knowing it, devotes himself' (1999b: 429). The *flâneur*'s aim is not to understand the labyrinthine city: it is to see it, feel it, and immerse himself in it. For Baudelaire, the poet/*flâneur* alone is capable of finding refuge in the crowd:

> Multitude, solitude: equal and interchangeable terms for the active and productive poet. The man who's unable to people his solitude is also unable to be alone in the busy crowd.
>
> The poet enjoys an incomparable privilege: in his own way he is able to be himself or someone else. Like those wandering souls in search of a body, he enters anyone's personality whenever he wants to. For him alone all is vacant; and if certain places seem closed, it is because in his eyes they are not worth the trouble to visit.
>
> The solitary, thoughtful stroller finds a strange intoxication in this universal communion. The man who easily joins a crowd knows feverish pleasures that the egoist, sealed up like a box, or the sluggard, closed as a clam, will always miss. He adopts as his own all the professions, all the joys, all the miseries which circumstances supply.
>
> (1991: 355)

But, as Benjamin cautions us, *flânerie* is not an activity that can be carried out in any city, at least not in the same way:

> The *flâneur* is the creation of Paris. The wonder is that it was not Rome. But perhaps in Rome even dreaming is forced to move along streets that are too well paved. And is not the city too full of temples, enclosed squares, and national shrines to be able to enter undivided into the dreams of the passer-by, along with every paving stone, every shop sign, every flight of steps, and every gateway? The great reminiscences, the historical *frissons* – these are all so much junk to the *flâneur*, who is happy to leave them to the tourist.
>
> (1999b: 265)

If Rome is too full of historical monuments to be a good stage for uninterested *flâneurie*, in Berlin the *flâneur* can easily 'depart from the ideal of the philosopher out for a stroll, and assume the features of the werewolf at large in the social jungle' (ibid.: 265).

In the contemporary megalopolis, concern about how to represent it may have become a predominant topic, as we saw in the remark made by García Canclini, but what relationship would this monstrous city have with fragmentary form? Is to represent speed, a well-recognized modernist ambition something aimed for by the artists of the megalopolis? And how does *flâneurie* fare in Mexico City and São Paulo? Perhaps like Rome's, Mexico's ancient monuments (constant reminders of both the triumphs of American civilization and the violence of the European invasion) will always somehow thwart idle *flâneurie*. In São Paulo, nothing is ever quite right for the nostalgic *flâneur* who returns to Vila Madalena from exile in Paris, in Hugo Giorgetti's film *O Príncipe* (2002). Meanwhile, in these megacities traffic, pollution, and violence cannot fail to take their toll of even the most intrepid of walkers.

Here we approach these issues by concentrating, first, on the use of fragmentary form in works by two young writers from São Paulo that have appeared in the past decade, Luiz Ruffato's *Eles eram muitos cavalos* (2001, They Were a Lot of Horses) and Bruno Zeni's *O fluxo silencioso das máquinas* (2002, The Silent Flow of the Machines), along with the blockbuster film made and set in Mexico City by the director González Iñárritu *Amores perros*, portending the millennium in the year 2000. What links these three works is their perception of time and the measuring of it in short periods, fragments of a single day (one of the Aristotelian unities) in the case of Ruffato's book, seconds or fragments of a second in the poetry of Bruno Zeni and in the film *Amores perros*. In all three cases, the fragmentation of time effectively becomes the means of conveying the inapprehensibility of space in the megalopolis.

This provides the context of a more detailed look at a few examples of *flânerie* – or of what it has become – in the two cities. In Mexico, these are the *flâneur* turned hired-killer in *Amores perros*; Carlos Monsiváis's immersion in the crowds in *Los rituales del caos*; Armando Ramírez's depiction of class and city-walking in the *crónica* 'Muerte anónima: el Rasguños' and 'El amigo de Catarino Vélez regresa al lugar de los hechos'; and Cristina Pacheco's televised peripatetic mosaic of Mexico City's poor *Aquí nos tocó vivir*. In São Paulo, they are the fatal walk taken by a young woman in the short film *Opressão*, and the gung-ho expedition undertaken by the Urbenauta Eduardo Emílio Feniano, *São Paulo: uma aventura radical*.

A DAY IN THE LIFE

Taking its title from a poem by Cecilia Meireles (who like him came from Minas Gerais), Luiz Ruffato's book *Eles Eram Muitos Cavalos* was published in 2002. Its author was then unknown on the Brazilian literary scene, although he was already the author of two books of short stories, *Histórias de Remorsos e Rancores* (1998), and *Os Sobreviventes* (2000). Both deal with his place of origin, the medium-sized town of Cataguases, in the state of Minas Gerais. By the time he came to publish *Eles Eram Muitos Cavalos*, Rufatto had had experience of São Paulo, living and working as a journalist.

'Book' rather than 'novel' or 'short-story collection' best describes Ruffato's work since, as Fanny Abramovich points out on the cover, the genre of *Eles eram muitos cavalos* is not easy to determine:

> I don't know whether I read a novel or a novella, short stories, an official report or a thriller . . . I know I hurled myself voraciously through the seventy flashes, takes, zooms in traversing the suffocation of São Paulo. It's a montage, a tremulous and mobile panel of people rushing to avert their gazes.

The seventy parts or sections of the book, numbered and titled, are connected to each other due to the fact of being set nominally in one place (likewise in line with the Aristotelian prescription of unities), the city of São Paulo, and due to the circumstance that they occur on the same day, 9 May 2000 (again the millennial year). The choice of this particular date had to do with changes in Ruffato's own life which had the effect of plunging him as it were naked into the minute-by-minute reality of the city.[3] Entitled 'Cabeçalho' (Heading)', the opening section invokes the endlessly recapitulated beginning of the school day, as it was experienced by children of Ruffato's generation, when they were taught to begin an exercise by writing the date and location, in this case identifying the place where the action occurs. Indeed, if we eliminate the titles and read the first three sections in sequence, we see that they also reproduce formulas used to introduce certain early morning radio programmes in the city of São Paulo:

> São Paulo, 9 May 2000. Tuesday. Today in the capital sky cover varies between cloudy and partly cloudy. Minimum temperature 14 degrees.

Maximum 23 degrees. Pollution Index, in the range of fair to good. The sun rises at 6.42 and sets at 17.27. Moon in the first quarter.

(2001: 11)

Immediately after comes a briefing on religious matters typical of some of these programmes.

By beginning with precise data followed by metereological and ecclesiastical information, Ruffato's book appeals to a circular notion of time, a time as it were cosmic (one day), underlined through repetition and opposed to the sense of novelty which normally characterized the stroll of the modernist *flâneur*.[4] Of course the notion of circular time has to be viewed with care, not just because it has served to rob those whom the West calls 'primitive' of their own sense of history but because (as Henri Lefébvre demonstrates in *Rhythmanalysis*) circular time and linear time never exist in isolation, independently one of the other, and because repetition (a concept that Lefébvre took from Deleuze, as we shall see later on), never reproduces anything that is exactly the same. In tying together the many sections of his book through the cosmic reality of a day, from sunrise until late at night, Ruffato creates the expectation of sameness, implying this is a day like any other in São Paulo. Yet he also manages to make us see this day as a fragment, as just one day among many in the overwhelming reality of the city of São Paulo.

Enjoined by Aristotle, taking the day as a time unit and unity as such was hardly an innovation. After World War I, the day lent coherence to one of the most famous modernist representations of the city, Walter Ruttmann's film *Symphonie einer Großstadt* (1927), which begins with Berlin waking up to a day of work and ends into the night. The film shows the masses moving through the modern city, and the individual drama of being lonely and alienated. The period of a day, a work day, is one of the basic units of Marx's *Das Kapital*, and informs many accounts of proletarian life. Recall, for example, the much-cited caricature of the factory worker's day offered by Charles Chaplin in *Modern Times*, where a man's sense of alienation is such that he begins to behave like a machine. The worker represented in this way forms part of an identical mass of workers in uniform who act in unison, like clockwork; indeed, *Modern Times* was also known under the title *Masses*, especially during its production. The metaphor most often used to describe these workers is precisely that of cogs in the machine, people who had lost their individuality, sacrificed to a system insatiably anxious to devour millions of workers like them, and their time.

In *Eles eram muitos cavalos*, descriptions of masses hardly exist. The idea of the city, of its enormity, is given through the multiplication of stories of individuals, stories different from one another which potentially could be multiplied to infinity, like the city itself. Nor are we allowed to observe the city through the eyes of a *flâneur*, eyes that would transmit fragments of passers-by, shop windows, lights, monuments and vehicles. The city is internalized not through the drama of one individual but by the tiny dramas of many individuals. The book is fragmentary, that is, made up of fragments, precisely because it points to the multiplication of itself.

The reader is made aware that the fragments or sections – seventy in all, like three-score and ten years of the proverbial life – are just a minimum among many further possible sections, since they occur in an urban area animated by approximately twenty million human lives. The seventy sections assume this awareness on the part of the reader: they exist in order to be just a tiny proportion of something much bigger. The theme of incompletion is announced throughout the book, in partially transcribed lists, truncated phrases, questions left hanging in the air, unanswered. 'Na ponta do dedo' (1; At your fingertips), for example, accesses the reader to an alphabetically classified list of jobs, that starts with 'Galvanizador' (galvanizer) and ends with 'Maçariqueiro' (welder. 41). It takes for granted that there are or were jobs listed under the several previous and the several subsequent letters of the alphabet, that the list provided is no more than the partial extract of a larger, indefinite whole. We are given lists of personal ads too, classified according to the category sought, husband, wife, friend, same-sex partner, even group sex. That is, the advertisements not only make clear that the lists are incomplete, they gesture towards further possible interpersonal relationships, like mirrors reflecting each other, once again, to infinity.

Besides classified advertisements, the book reproduces notes, menus, inventories as of books on a shelf, a tally of objects in an empty flat, horoscopes, self-help texts and prayers, like one to Santo Expedito. In smaller type, this last gives information about where it was printed and explains that a thousand copies are being distributed as the means of paying for a wish that had been divinely granted, so that in this way the prayer is, as it were, broadcast through the city. In the words of Marguerite Harrison, these texts 'reproduce the multiplicity – the bombardment – of information meant to enhance communication in a global age' (2005: 157).

Most of the seventy sections, meanwhile, might be considered 'sketches' in an almost theatrical sense. They are small excerpts, minimal dramatic situations which allow the observer to get a sense of a conflict or of a narrative sequence that extends beyond the covers and limits of the book. Yet written down here they are 'dramatizations' that almost always involve more than one character, multiplying themselves, again, indefinitely. Although many of the sketches are narrated by an omniscient third person, the narrative voice seldom strays beyond giving descriptions reminiscent of stage directions, with respect to scenery, gesture, or costume. When the voice does go further, it underscores the point of view of the characters in free indirect speech, and there is plentiful recourse to dialogue and monologue. These qualities of the work doubtless led to its being promptly converted into an actual play, entitled *Mire Veja* (Look See), which was performed to acclaim in a central theatre.[5] Curiously, even the daring graphic quality of *Eles eram muitos cavalos*, which Harrison has analysed in detail, seems to function as a kind of 'choreography', since it locates the dialogues on the page as if they were characters on the stage, and it 'dresses' differing kinds of speech in corresponding fonts (italic, bold) as if to personalize them. This typographical experimentation attempts to compensate for the linearity to which alphabetic script is normally limited, creating the order of simultaneity more readily found in theatre, thanks to positionings on the stage, for example, or the simultaneous use of various forms of communication (that is, two or more characters may talk at the same time, while images or texts are being projected or music is played, and so on). In *Eles eram muitos cavalos*, a telling use of graphic resources occurs in fragment number 16, which is set in a helicopter flying over the city:

> – i'm not insensitive to the social question *unrecognizable the city centre hordes of hawkers pickpockets sandwichboard men reek of urine reek of saturated oil reek of* a hand strokes her thin hair **(my mother was putting on her gloves, hat, big jump to get across the viaduto do chá** [at the centre of the city], **me, a boy, really tiny, was running on).**
>
> (2001: 36–7; Author's emphasis)

In many of the fragments, the interrupted sentences indicate speech-acts aborted or unfinished, incomplete thoughts, or maybe a listener who had got distracted and stopped paying attention. They evoke, one way or another, the sounds of the city which we hear only partially, which float through the streets to be heard, who knows, by someone or other.

This sensation of incompleteness, of parts that imply further parts, multiplying to form not a whole but an ever provisional combination of many parts, is what most immediately helps us define the genre of *Eles eram muitos cavalos*. As Abramovich noted in the blurb, we do not know whether the sketches configure a novel, a novella, or a collection of short stories. Strange as it might seem, this last suggestion is the least likely. In spite of there being no narrative thread to tie the sections together one to another, nor characters shared in common, nor over-arching plots, the book between its covers conveys a strong sensation that the sketches should be read as a whole, that is, that they can be better understood in connection one with another. They can of course be read separately as well, and some of them are so dense as to be able to survive were they to be included, for example, in an anthology of short stories. Yet there exists a coherence between the sections that is due in the first instance to space, that is, to the fact of their taking place in São Paulo. In the words of Harrison: 'In fact, on a structural level, it is the very presence of the city that distinguishes this work of fiction as a novel' (2005: 151).[6] But that, yet again, is problematic. The dramas that make up the sketches take place in different parts of the city, so different in fact that the mere idea that they could all belong to 'the same place' cannot fail to seem odd. For although some of the sketches take place in truly 'iconic' parts of the city (the corner of Rebouças and Brasil Avenues, for example, or Ibirapuera Park, or even certain streets in the old centre like 24 de Maio), most happen in nameless neighbourhoods, houses of the poor or condominiums of the rich, on anonymous streets. The characters involved never cross each others' paths and never relate to each other. Every one perceives the city in her/his own way, but it is the city that in one way or another determines the rhythms of their lives. The city emerges as a concept at once omnipresent and vague, impossible to define, no less impossible to ignore.

Practically no proletarians live in Ruffato's São Paulo, and the reader never has the sensation, so frequent in modernist writing, that the workers are a mass of beings identical to each other, cogs in the machine. Indeed, there are few workers of any kind in the book, which is haunted by the spectre of unemployment. In one of the sketches, the reader accompanies a lower middle-class lad as he gets ready for a job interview, his tenth in two months. Later on, his mother is watching the news alone in the house, worried because her son has not come back. In another, a beggar is keenly observing a residential building, which causes those who live there and their employees to feel threatened. We find out later that he had worked as a janitor in that same block, and that he had lost his job after a personal loss that turned him into an alcoholic and led him to quarrel with a resident. In 'Trabalho' (Work), we witness a young man growing tired of working for little money, in what he calls a system of consensual slavery; taking his wife and children with him, he goes off to live with his parents-in-law. In 'Brabeza' (Pluck), a young man roams the city centre, in search of possible victims to rob: Mother's Day is approaching and he wants to give his mother, who is ill, a present. He doesn't take drugs and doesn't consider himself a thief. He is robbing people now but dreams of one day getting a good job. In 'Fraldas' (nappies), 'a giant black man, broad-shouldered, impeccable in his black suit' who works as a security guard in a supermarket glimpses 'a frail black man, boney, clad in a filthy check shirt, dirty jeans, worn-down tennis shoes' (2001: 54) who is about to steal some baby items. When confronted, the would-be thief confesses that he had wanted to 'borrow' the items in question because he had lost his job and his wife had just given birth. And more in the same vein. Far from being posed as an abhorrent situation, unemployment appears in the book as the state of near normality, while conversely the dream of finding a good job, ever unrealizable, acquires the airs of a chimera.

Examples abound of people whom Marx had assigned to the lumpen-proletarian realm of rags. A drunk killed probably by the police – what we learn about him is transmitted by a dog, which carries on hunting him down. A woman whose small children sleep in a bed full of excrement, insects, and rats. A prostitute about to score. A hairdresser who caught the AIDS virus from her unfaithful and violent husband and now lies dying abandoned and in penury. Plus the new lumpenproletariat: teachers who barely earn enough to survive and who have to face the violent reality of the schools where they work and the neighbourhoods where they live.

But violence is not just a problem of the poor. It is everywhere and determines no less the day-to-day life of the rich. In the emergency room, for example, a doctor refuses to operate on a patient when he recognizes him as one of the gang that had threatened to assault his family. The occupant of a luxury condominium imagines how he could have been the friend of a neighbour who has just been murdered, the victim of a 'lightning' kidnap (whose credit cards are put to instant felonious use). A man driving around the corner of Rebouças and Brasil in a Mercedes that is heavily armoured. Few sketches portray violence as such. For the most part, what we have is fear, greater and lesser, will and frustration, and an all but permanent sensation of insecurity. In the last sketch in the book, a couple is awakened in the middle of the night by groans that seem to come from the other side of the door, and they are afraid to go and find out what is going on. Like them, we, the readers, end the book not knowing the source of the groaning, not knowing whether the violence they fear is real or imaginary.

Speed is absent in this portrait of São Paulo. Indeed, traffic is constantly at a standstill or very slow, and people interact personally by telephone or the internet inside their houses, bars, restaurants, schools, hotels, and in taxis in perpetual search for ways around the snarl-ups, or up in helicopters (often used by the very rich to avoid traffic or for security reasons), and making their way through streets crowded with people and armies of street sellers. In part, this corresponds to a realist portrayal of a city where traffic problems are in fact an everyday reality. Yet the seizing up of the traffic and the snail's pace that constantly paralyze the city in *Eles eram muitos cavalos* point no less to a stagnation that is social. The chasm that separates rich from poor might be said to be a main theme of the book, like the almost non-existent upward mobility, most heavily marked among the lowermost classes.

Nor does the book allow room for the attitude of the *flâneur* who makes his own way through the city. Instead of a single man walking through São Paulo, we see hundreds, thousands or millions of men, women, children, animals and objects moving along, never meeting, never registering each other's existence. This absence of encounter generates as it were the space between the sketches, which in turn becomes the main narrative line of the book. In other words, what paradoxically makes a novel out of *Eles eram muitos cavalos* is the eloquent lack of an integral plot, a lack which is not exactly absence, but rather the presence of negation. Hence, the city is defined not by a readily recognizable central image but by the multiplication of parts that are unknowing and ignorant of each other. And these parts are at one and the same time both similar to each other and absolutely different. At opposite poles to an unequivocal, univocal portrait of the city, the sketches are marked by repetition in the Deleuzian sense, that is, as a negation of sameness or identity.

In tracing the genealogy of his philosophy of repetition to Nietzsche, Kierkegaard and Péguy, Deleuze explains that, allowing for all the differences between them, these thinkers had promoted 'a prodigious encounter around the philosophy of repetition: they oppose repetition to all the forms of generality' (1994: 13). The modernist portrayal of the masses is ostensibly an attempt to achieve generality, that is, a conceptual definition of the 'worker', the 'proletarian', the 'poor'. For its part, the modern megalopolis that emerges in *Eles eram muitos cavalos* points rather to the lack of generic conceptualization: there is no univocal portrait of the city, a central idea of what São Paulo might be. The city is defined precisely by this lack of conceptualization, by the multiplication of different parts each with its own centre but which are superimposed or happen in a rhythm, the rhythm of repetition. As Deleuze notes: 'There is a big difference between generality, which always designates a logical power of the concept, and repetition, which is testimony to the lack of power or of the real limits of concepts' (ibid.: 22).

For good reason the book is divided into similar sections or sketches, that is, into 'dramatizations' of specific situations. For Nietzsche, as for Deleuze, dramatization diametrically opposes abstract conceptualization, as can be seen in the critique that the latter makes of Hegel: '[Hegel] represents concepts as opposed to dramatizing ideas' (ibid.: 18). The sketches in *Eles eram muitos cavalos* are 'dramatizations of ideas' in just this sense, live situations that imply a plurality of nodes, a superimposition of perspectives, a tangle of view-points. For Deleuze, truly modern art (like Joyce's *Finnegan's Wake* or Mallarmé's poetry) not only multiplies points of view:

> Each perspective or point of view should encompass a work in a self-sufficient sense: what matters is divergence in the series, a de-centering of circles, monstrosity. The totality of circles in this case is a formless, rootless chaos, with no law other than the repetition and the reproduction of itself, and from there the developing of that which diverges and decenters.
>
> (ibid.: 94)

This quotation could well describe the movement of repetition created by the sketches in *Eles eram muitos cavalos*, for through them there emerges precisely a 'formless, rootless chaos', that is, the city as monster.

Repetition sets up rhythm, and rhythm is another characteristic that is explicit in Ruffato's book. Many of the sketches include internal repetitions of phrases, songs, tics. Fragment number 4, for example, the first of the 'dramatic' sketches, is punctuated by the tum-tum tum-tum of the music playing like a heart-beat on the car radio of the protagonist, a man who is driving to the airport. It is punctuated, too, by the song's refrain, set in italics: 'mais neguim pra se foder' (2001: 11; One more for getting screwed). By these means the rhythm of the music sets up another rhythm, that of birth ('mais neguim') and of sex, suffering and death ('se foder'), that is, of the human life whose years correspond to the number of sections that make up the book. The refrain further refers us to a multiplicity of meanings: 'neguim' may just as well denote anybody or any person as it can, when taken literally, specify someone who is black, establishing at the very start of the book the theme of racial prejudice, always so covert and so ambiguous in Brazil, and the consequently miserable prospects in life that await the children who are born poor, and black. This line of thought is set beside the life story of the anonymous character who is driving the car. His success as the agent of a corrupt businessman does not atone for the fact he himself is a subaltern employee (um empregadinho) who is never going to be dealing on an equal footing with the boss's daughter. The first sketch prefigures then the rhythm of the book, and counterposes a possible notion of indistinguishable masses ('mais neguim pra se foder') with that of differences characteristic of individual dramas.

'Pelo Telefone' (By telephone) repeats in successive calls the message on an answering machine: 'Oi, aqui é a Luciana. Deixe o seu recado após o sinal' (Hi, this is Luciana. Leave your message after the tone; 2001: 52). In each one of the calls, a woman leaves an offensive message, so that bit by bit a triangular relationship is revealed involving the lover (Luciana), the wife (who leaves the messages) and an unfaithful husband. The exact repetition of the answering machine message contrasts with the messages left on it, apparently much alike, but which gradually build into a spiral where what initially seemed to be devotion to the unfaithful husband becomes resentment, contempt, and even disgust.

The cycle of repetitions in *Eles eram muitos cavalos*, stretches through the cosmic time of one day, the book's most palpable unit and unity. Yet even the day is a precarious unit since as we already saw the immense variation in experiences and their incompleteness insinuates that the 'day', like Nietzsche's eternal return, the coming back that never ends (*ewige Wiederkunft*), brings more difference than similarity. 'Through eternal return, Nietzsche meant nothing else. Eternal return cannot signify the return of the Same, because it presupposes, on the contrary, a world (that of the lust for power) where all prior identities have been abolished or dissolved' (Deleuze 1994: 59).

In this fashion, recourse to the fragment in *Eles eram muitos cavalos* privileges repetition, the source of difference in identity. If, as we saw, for Deleuze that would be a diagnostic of the radically modern work of art, what to say about *Eles eram muitos cavalos*, which on a first reading would appear not to demand the same sense of rupture with language, nor to appeal to a vanguard radical to that degree? In fact, it could be argued that the obviously social concern of *Eles eram muitos cavalos* would sooner tie it back, up to a certain point, into the realist tradition. Harrison has demonstrated, however, that formal experimentation is central of the aim and purpose of the book:

> Unlike other contemporary writers who might represent a present-day reality through straightforward narratological constructs, Ruffato insists on a creative path that juxtaposes true-to-life representation of urban reality with striking literary experimentation. This attention to formal innovation does not clash with the novel's social content; rather, is designed to both underscore and coalesce with reality, even to the extent of exacerbating its inclemency.
>
> (2001: 155)

The absence of a univocal centre and the recourse to a repetition that establishes difference help us realize that, as Harrison says, formal experimentation and ethical concern are indissociable in *Eles Eram muitos Cavalos*. Or who knows, perhaps this provisional portrait, made up of incomplete parts, mirrored, incommunicable and multiplying each other to infinity, in the end might be the only viable realism for the São Paulo of our time.

MACHINE FLOW

Rhythm is also at the heart of Bruno Zeni's *O Fluxo Silencioso das Máquinas* (The Silent Flow of the Machines), published the same year as *Eles Eram muitos Cavalos* and likewise set in the millennial year 2000. It was Zeni's first book, although like Ruffato he also had had experience of working in journalism. Since then, Zeni has co-authored a book by a former prisoner (*O Sobrevivente André du Rap*), and has written articles on *paulista* rap.

Not instantly apparent, the genre of *O Fluxo Silencioso das Máquinas* might best be defined, again, as fragments, this time of poetic prose. The sentences are not generally lined out as verse – i.e., they follow the linear conventions of prose – yet in being gathered in stanza-like fragments they tend to be positioned along the bottom of the page, below Cagean space or silence, emptiness in which they can resound. Many of the pieces name the city of São Paulo, while others refer to it directly or indirectly, so much so indeed that even those pieces that do not refer to the city become, by implication and association, connected to it. Like *Eles Eram Muitos Cavalos*, *O Fluxo* is a book about and of São Paulo. Once again, it is the city that links and draws the fragments together, even though the links are somewhat tenuous, more so than in Ruffato's book.

Each fragment in Ruffato's book deals with different characters, and the fact that one character from one fragment relates to none in any other fragment becomes, in itself, the main connection between them, according the kind of negative definition we have been examining. In other words, in terms of reader expectation the lack of connection between the characters becomes a model, a structural device that makes the reader expect and positively anticipate a different set of characters, from one fragment to the next. For its part, *O Fluxo Silencioso* creates no such expectation from one fragment or poem to the next. While some pieces seem to coincide in reiterating a point of view or a particular take on the city, others do not, standing free and autonomous, like poems, the term used from now on to refer to the constituent fragments of the book. While some poems seem to share the same poetic 'I' – a particular voice that observes the city and its people – others sooner suggest the order of bric-a-brac or collage found in Ruffato, being structured as transcriptions of newspaper articles, news programmes, snippets of popular science (possibly mediated by the press), horoscopes, etc. What links them, what makes them coherent as parts of a possible whole is, once again, the city of São Paulo. Unlike *Eles Eram Muitos Cavalos*, however, *O Fluxo* makes no attempt to represent, embrace or conjugate the many very different parts of this city, or the *periferia* regions. Most identifiable references are to relatively central and iconic features, large avenues like Rebouças and Faria Lima, or 9 de julho, for instance.

Zeni's frequent references to machines are reminiscent of the modernist literature about cities, but here they are, paradoxically, centred on the image of a machine that does not move, clockwork whose gyrations around itself make it go nowhere. As a note at the end of the book explains, one of the pieces is based on a programme transmitted by the elitist Eldorado radio, which insistently urges its listeners: 'Call us here and tell us where you are and where you are bound in São Paulo' (2001: 61), alternating it with practical updates on traffic. These normally will confirm that, as usual, the traffic is stuck in a jam: 'At the moment, tailbacks in the city are 182 kilometres long. Avenida Paulista is still not moving in either direction' (ibid.: 61), or 'The traffic is at a complete standstill along practically the whole length of Tietê bypass' (ibid.: 62).

In the first poem, a homeless man is described in the third person as if he were on a beach, his head resting on his small bundle of belongings, observing the 'tide of cars' go by that gets to be 'full by eleven at night'. In an introductory fragment that precedes the table of contents, the city is described as 'lined with cars', which are then compared to the red sunset of the polluted sky: 'The cars also grow red at the end of the day, with the brake lights that flicker and glimmer. Stopped, they pant. Breathing in and breathing out, just them, the only ones to be drawing breath' (n.p.). While cars in modernist representations of the city tended to emphasize the fragmented view given by speed, in Zeni, as in Ruffato, they indicate stagnation, the heart-beat rhythm of a city barely staying alive, in suspended animation, medically monitored. Of course the frequent reference to traffic congestion is to some extent again a realist element: traffic is indeed deadly slow and sluggish in São Paulo. But it also stands for a more general sense of stagnation, a seizing up that is the reverse of the promises of a better future found in both the fascist celebration of technology in Marinetti's famous futurist Manifesto of 1909 (very influential in the São Paulo *modernista* movement) and in revolutionary novels like Patrícia Galvão's *Parque Industrial*. In Zeni, the homeless man of the first poem is simply there, part of a cityscape that he shares with the futuristic *telão*, one of the two or three giant electronic screens recently set up in São Paulo along the main traffic arteries which, whether they like it or not, shower commercial propaganda, official announcements and even poetry on to a public that most of the time is calculated to be literally captive, i.e. the occupants of motionless vehicles. From his vantage point, the reclining tramp distinguishes only 'gold, reds, greens, blues and purples, all mixed up, blurred and liquid' (2001: 15).

The congested traffic emblemizes a larger notion of urban stagnation and seizure, so overpowering that no-one can move through it or escape from it. 'There is no air. Here everything is big, but it is hard to move' (ibid.: 10). In the third person, one poem focuses on the desperation of a man waiting at a bus stop who gives the impression of being about to crack up or have a fit. 'He does not move, but his arms clutch his folded legs. He does not move, but he is on the verge of an attack' (ibid.: 35). The attack itself is marked by the 'near screams' that he utters, which are described as 'marking a rhythm that will keep his thoughts from shattering' (ibid.: 35).

The disjuncture between internal rhythms and sounds born from individuals, and the external rhythms and sounds of the city is a constant theme throughout the book, one reminiscent again of modernist texts. In one poem/fragment, a person (the reader cannot tell whether it is a man or a woman) goes down the metro escalators listening to music (probably on a portable device), observing the other travellers and how they (and external objects) are endlessly reflected on the train windows. The poem plays with the artificial rhythms that rule timetables and life in the underground and the lights it reflects, counterposing them to the personal rhythm of this undefined first person. He/she walks in time to the music coming in through the ears which, it is said, 'makes you go really wild' (ibid.: 24), a state clearly at odds with place the body occupies, entrapped if not entombed in the underground. Another piece describes a middle-class teenage boy who walks 'as if to the sound of guitar bass and drums, with an imaginary gun drawn in his hand' and 'planning repeatedly crazy stuff for himself' (ibid.: 41). A poem, entitled 'Tecno', lists metallic sounds that test at the 'limits of bearable': a turbine, electric needles, the sound of a modem, and 'pressurized samba'. These sounds create 'A new language, a bit harsh, a bit stupid, a by-product of the sonic humming', and are 'reverberations repeated like a mantra. We're alive' (ibid.: 63). On the one hand, the poem celebrates, in futuristic fashion, machine sounds. On the other, it laments those sounds as 'stupid' language and ironically comments on the fact that nowadays it is only through machines that we can know whether or not we are alive.

Rather than speed, many poems attempt to simulate the invisible waves that float and flow through the city, radio, television, light, and telephone waves, x rays, sun plasma, etc., as well as thicker versions in underground rivers and sewage. Invisible to the eyes and other senses as waves, the former generally rely on the language of physics to be comprehended or imagined, in practice most often for most people what is digested in a brand of popular science that is itself conveyed by radio waves:

> Since our equipment for perceiving waves is very limited, we do not notice the radio waves that go through us (the opposite of light, visible to the eyes, and heat, which we feel on the skin). The same happens to other kinds of electromagnetic waves with even bigger frequency than visible light, like ultra-violet radiation, x-rays and gamma-rays ('and some can be very bad for the health').
>
> (ibid.: 45)

Yet, these waves penetrate our bodies in the city, through a process that Paul Virilio has referred to as a 'new scientific definition of surface' which 'demonstrates the contamination at work: the "boundary, or limiting surface" has turned into an osmotic membrane, like a blotting pad'. In other words, for Virilio (and for Zeni), in the bombardment and the radiation of the human body characteristic of the modern city, a new spatial definition is at work:

> What used to be the boundary of a material, its 'terminus', has become an entryway hidden in the most imperceptible entity. From here on, the appearance of surfaces and superficies conceals a secret transparency, a thickness without thickness, a volume without volume, an imperceptible quantity.
>
> (1991: 90)

The key to this definition of reality in Zeni's book is its title poem 'O Fluxo Silencioso das máquinas'. In the original, it begins:

> O fluxo silencioso das máquinas. Diversos prédios possuem em seu topo pequenos focos puntiformes de luz vermelha. A balizar as rotas de avião. De cima, a cidade é um campo de luzes amarelas, brancas e vermelhas. Um mar de luzes que piscam. Já alguns prédios, quando se passa por eles, acendem um grande facho branco. A pessoa vem andando pela calçada e quando passa pelo sensor – geralmente instalado na guarita do porteiro – faz-se a luz. É um susto, quase sempre. Um mergulho tembém, já que não se enxerga nada por uns momentos.
>
> (2001: 101)[7]

The opening definition derives from the electric and electronic impact on senses themselves mechanized, which then blind and disorient those of the human, machines and lights and that work all night: the 'silent flow of the machines'. They include the red lights on top of tall buildings that guide airplanes; movement-sensitive security lights. Blinding, these lights plunge the person as it were into the depths like a diver. Consciousness recovers through the unvarying instructions of automated teller machines (welcome. insert your card. enter your pin number), all in 'the sea of buildings that do not sleep' where there are

> computers that stay up all night. Where do their waves go? Confirm your password. You are entering a security area. Do you want to continue? The information you enter will be cryptographed for your safety. The discreet humming of air-conditioning. The thick smell of the environments. Which press the lungs one against the other, making them close. A token for the drink machine, please. The thumping fall of a soda can. Enter your six-digit password one more time. Insert and remove your card to release the bills.
>
> (ibid.: 101)

It is Virilio, again, who gives us a definition of urban time that much resembles this poem by Zeni:

> Where once the opening of city gates announced the alternating progression of days and nights, now we awaken to the opening of shutters and televisions. The day has been changed. A new day has been added to the astronomers' solar day, to the flickering day of candles, to the electric light. It is an electronic false-day, and it appears on a calendar of information 'commutations' that has absolutely no relationship whatsoever to real time. Chronological and historical time, time that passes, is replaced by a time that exposes itself instantaneously. (. . .) As a unity of place without any unity of time, the City has disappeared into the heterogeneity of that regime comprised of the temporality of advanced technologies. The urban figure is no longer designated by a dividing line that separates here from there. Instead, it has become a computerized timetable.
>
> (1991: 88)

Hence the primacy of the moment, or the *instante* in Zeni's poetry. In 'O Instante quer fugir' (The instant/moment wants to escape) the poet starts by defining the *instante* in broad, relativistic terms: 'It is typical of the moment. This thing of escaping. This thing of the moment can last a long time. Because if we are talking years, the moment can be one day. It can also disappear one day. One day, who knows. What is left of a step after the other?' (2001: 29). He then moves to the specific seasonal and threatening temporality of the city that animates his book. 'It rains a lot in São Paulo in February, March. There has even been someone who shipwrecked in Anhangabaú, who got washed into the tunnel, dragged by the current' (ibid.: 29). The circular seasonal time is then blended with the linear time of the *instante* that moves away, cancelling out (once again, as in Ruffato), any possible difference between them: 'The season of rains: each one defends himself as he can. Under the marquee, under umbrellas, wrapped up in capes and hoods, inside cars, behind windows – one needs a shield to go on, step over step in the moment' (2001: 29).

In 'Transexual. Internet shows sex-change surgery', the *instante* described is that of the simultaneity and timelessness of new technology. In language that repeats that of news broadcast, we are informed of the 'first sex-change surgery with live broadcast in the internet'. The promise of global accessibility implied by the internet is undermined by the fact that 'out of the 14 thousand people who entered the site, 821 managed live access and could see the operation' due to an 'excess of consultations' (Zeni 2001: 91). However, 'The images are still in the air' – in an incongruous ever-presentness of a moment that should represent radical personal change. This new possibility of timing is described, as well, by Virilio: 'The greatest geophysical expanse contracts as it becomes more concentrated. In the interface of the screen, everything is always already there, offered to the view in the immediacy of an instantaneous transmission' (1991: 91). It represents, in his view, a switch from 'the extensive time of history to the intensive time of momentariness without history – with the aide of contemporary technologies' (ibid.: 119). For him, this new time is, in all accounts, negative or false. In *The Last Vehicle*, he goes so far as to ask his readers not to trust this new sense of time:

> Let us not trust it. The third dimension is no longer the measure of expansion: relief, no longer the reality. From now on the latter is concealed in the flatness of pictures, the transferred representations. It conditions the return to the house's state of siege, to the cadaver-like inertia of the interactive dwelling, this residential cell that has left the extension of the habitat behind it and whose most important piece of furniture is the *seat* (*siège*), the ergonomic armchair of the handicapped's motor, and – who knows? – the bed, a canopy bed for the infirm voyeur, a divan for being dreamt without dreaming, a bench for being circulated without circulating.
>
> (Virilio 1989: 120)

In Zeni's poetry, by contrast, the permanent presentness of the internet, however incongruous or absurd, still bears the possibility of liberation and change (not excluding sex change). Neither does its simultaneity occlude political difference: In 'Passeata gay leva 100 mil à Paulista' (Gay parade brings a hundred thousand to Paulista Avenue), the structure, once again, of a news text informs the reader of two events: the gay parade in São Paulo, presented with colorful details of the different characters and sexual choices, and with the information that it happened without incidents, and the gay parade planned in Rome, which is meeting with protests from the Catholic Church and right-wing activists.

Above all, the ever present simultaneity of new technologies does not seem to point to the house-bound siege of Virilio's text, as 'Se isso então aquilo. Ou não' (If this then that. Or not) celebrates the *instante* precisely as enjoyment of urban life or, in Nietzschean fashion, as the only possibility of infinite happiness or pain. First, the poem lists some simple moments of pleasure in being in the city: 'Walking in the city bodies together, looking at the profile of the buildings and bridges. The drawings. Graffiti. Colour. If possible play slot machines. If on a rainy afternoon, cinema' (Zeni 2001: 51). Then, it evokes the erotic pleasures of nocturnal urban encounters:

> At night, the dance floor and the lights, everybody dancing, very close, as if in the wind, max volume. If nipples hard against the blouse. If a dick hard in the pants. If a cigarette, if a drink also long. If a sweaty and dirty fuck. If all this can happen then the state of absolute and infinite happiness is possible.
>
> (ibid.: 51)

This poem, like others in the book, expresses the desires of an individ-ualized poetic voice. Although we are hardly presented with images of the poet walking in the city, we can easily imagine the poetic voice, in several poems, as that of a *flâneur* who observes urban landscape in flashes. It is the case, for instance, of the poem 'Pequena iluminação asfáltica (o menino vaga-lume' 93), which describes, in almost fluorescent colourful tones, a group of poor boys playing football on the street, at the end of a cul-de-sac. Playing with concepts inseparable from an idea of movement in Brazilian football, it ends:

> Shiny green, really green, scintillates as night falls.
> The boy's supple twisting [ginga de lá-e-cá] makes designs in the dark.
> The firefly boy is playing ball on the asphalt of São Paulo.
> The other boys gather around [fazem coro] and dance with him.
>
> They exchange passes in the cul-de-sac.
>
> > (ibid.: 93)

Clearly, the poetic voice is speaking and observing all this as an outsider, almost as a painter of the poetry of the street. Poems like this, are, in many senses, closer to a Baudelairean perception of the city than to the multi-centred fragments in Ruffato's book.

At the same time, this individualized city is also constantly transformed, and produced, by anonymous human presence. The poem 'The Writings. They Come up', describes the graffiti scribbled on the wall as

> belonging to nobody . . . The writings. They come up. Some say they are done by people who have nothing to do and there isn't really much to do around here. But they haven't discovered yet the why and the how. // The rain running down the black of the walls does not wash it off, just composes it. The static city in the calm of the spotless odorless concrete. // The mute and dirty concrete. Porous.
>
> > (ibid.: 37)

That is, graffiti, in the poem, permanently 'composes' the city, transforms it into an artwork that is constantly being modified by human presence; turns its spotless and hard concrete into porous, changeable surfaces.

Moreover, when we read the book as a unity, and as a portrait of São Paulo, we can see that the juxtaposition of discourses modifies the individual-ized perceptions of the possible *flâneur*. Horoscope predictions, for instance, appear twice – and their lack of context makes them into ready-made objects, as it were, as items to be consumed in the city. There are also many news pieces of different kinds and nature. One of them announces that

> The girl Tatiana Oliveira de Castro, 3 years old, died yesterday after having received a gun shot during a party in the municipal school

Synésio Rocha, in Jardim Umarizal, Campo Limpo (south side of São Paulo). Her abdomen was perforated by the bullet.

(ibid.: 39)

The event, according to the poem/news happened when a party was invaded by six men who fired at Alexandre Marques, who was hit by six bullets. Several pages later, another poem corrects the news, by saying that 'Differently from what was reported, the girl Tatiana Oliveira de Castro, three years old, has not died. She actually had her guts perforated by a shot during the party, but she was quickly taken to hospital and is in good condition' (ibid.: 49). It also corrects the other facts in the previous announcement by saying that Alexandre Marques was hit by four bullets, not six. Both the news and the corrections create a sense of the city that is highly contradictory, even nonsensical: on the one hand the 'corrections' point to the fact that news about violence in the city is exaggerated. On the other hand, although they make a serious claim in the case of the girl (she is not dead), they do not change the fact of the violence *per se*, even more so because the 'corrections' about Alexandre (four perforations, not six) will seem absolutely irrelevant and pointless to the ordinary reader.

The fragment poems that constitute Zeni's book make multiple use of received text, the news report (as here) no less than the radio programme, music, and scientific or pseudo-scientific discourse, suffusing it in the instant and the timelessness of new technology. All this subtly modifies the individual poetic voice, which in itself is accorded a strong presence, and re-defining the spaces occupied by the human body in São Paulo and its inner being in the megalopolis. As in Ruffato's work, it meanwhile amounts to a diagnosis of the megalopolitan body itself, perilously bloated and over-extended, as it were through anatomical function: clogged traffic arteries, suffocated lungs, stagnant liquid flow, and the seizing up of the motor heart.

In pointing to the same moment on the millennial threshold, Zeni's *O Fluxo silencioso* and Ruffato's *Eram muitos cavalos* deserve special attention in the sea of writings that have attempted to grapple with the megalopolis in recent years. Reaching similar perceptions from quite different starting points, which in itself might be thought to enhance their joint worth, these works acknowledge (not for a second, underestimating them) the possible proportions of the monster while responding with supreme finesse to the predicament and pathos in the lives of those who live in it, the lad going off to his tenth job interview in two months or the roadside tramp gazing at the *telão*. In form, their common appeal to the fragment, be it prose or verse, sharpens the reader's sense of the sheer material fact of the book and the page, and it opens up concern for how as a gathering the fragments best fit together or whether they can cohere at all. Technically reinforcing each other, both books ingeniously refract language and discourse that arrives already made, tending the lyrical voice within. All this enables or encourages the reader radically to question such rudimentary concepts as setting, story, plot, person, character, and the temporal realities in which these may exist, their identity, sameness and difference. In each case, the result amounts to an acute diagnosis of the megalopolis, a thorough pondering of its possible apprehensibility. Much of what may be learned from these verbal constructions of São Paulo in 2000 resonates in turn in cinematic language, in a major film that appeared in Mexico City in that same millennial year.

COLLISION

The DVD menu screen of the film *Amores perros* (2000) shows the stylized profile (in black) of a city, against a red background. The same image appears on the DVD cover. There seems to be no question, to judge from this visual evidence, that the city plays a central role in the film. The director Alejandro González Iñárritu confirms what the DVD cover and the first screen indicate:

> Mexico City is an anthropological experiment and I feel part of it. I am just one of the twenty-one million people who live in the largest and most populated city in the world. No man in the past lived (or rather, survived) before in a city with such levels of pollution, violence, and corruption. Yet, it is an incredible city and paradoxically beautiful and fascinating. This is *Amores perros*: fruit of this contradiction.
>
> (Eseverri n.p.)

Yet, as Paul Julian Smith observed, the film presents very few iconic images of Mexico City:

> It is hardly surprising that *Amores perros* avoids tourist shots of the Zócalo and the Historic Centre. But it also neglects the soaring and sometimes distinctive skyscrapers of the several business districts. This is curious, because the combination of ancient and modern is (uniquely for an American capital), the distinctive characteristic of Mexico City where dilapidated sixteenth-century chapels can rub shoulders with gleaming glass towers.
>
> (2003: 51)

While some might quibble about how 'unique' this juxtaposition of ancient and modern is in the American continent (or about identifying the ancient in Mexico with chapels rather than pyramids), Smith is characteristically perceptive in noticing how the film eschews the obviously iconic. For most of it is set in what apparently are non-descript places in the city: poor neighbourhoods, rich or middle class streets and avenues, houses and apartments from which the viewer can sense urban movement, without necessarily knowing where it is going, and so on. Very few images of the city allow prompt identification. If the film, as its director says, is about 'the largest city in the world', in which ways is the city itself allowed to emerge in it?

The very few architectural identifiers of Mexico City that do exist in the film are strategically placed, and quite crucial to the development of the story. At the very start of the film, a hand-held camera records a car chase which provides benchmarks for the story as a whole. What the viewer is allowed to see is seen from inside a car that is being pursued at great speed. It is being driven by Octavio (Gael García Bernal), while on the back seat Jorge (Humberto Busto) is attempting to staunch blood from the wounds of a dog, called Cofi. Against a background noise of cars (tyres screeching, horns, cars accelerating and overtaking), we get the briefest glimpse of the Torre Latinoamericana. One of the best-known landmarks on Mexico City's skyline for half a century, it may be recognized here as a swift, tremulous image. Economical as it is, this image is enough to identify unequivocally, right in the first scene, the place where the story is going to be set. In addition, the way it is presented – as a tremulous fleeting image, supposedly seen through a car window – this landmark of modern, globalized Mexico City, bears an uncertain, precarious, almost dream-like relationship to the two young men, who have got into trouble thanks to dealings with an underground world of violence.

Later on, the setting is confirmed by a few shots of the volcanoes and mountains that surround the city, landmarks invoked over the centuries and seen in Mexican books already before the European invasion (the screenfold Codex Vindobonensis, for instance, named after the place where it is now housed). Here the orological reference reminds us, as it were, of where we are and of the ancient rim of what is one of the biggest cities in the world, if not the biggest. Audially, the location will be clear to any local cinema-goer from the start, in dialogues that by no means censor the accent, vocabulary and turns of phrase typical of its *chilango* (i.e. Mexico City) population.

Amores perros was the first feature film by González Iñárritu, a successful DJ and director of advertising films. It became a major hit, especially after it won the Cannes Critics' Week award, and was nominated as best foreign film in the Golden Globes and Academy Award. It has been watched by very large audiences not only in its native Mexico, but also in Europe and the United States. A quick browse on the internet will pull up hundreds of reviews, in several languages. Positive reviewers often mention its clever tripartite structure, brilliant camera work (the film makes ample use of hand-held camera) and astounding pace. A few negative reviews point to its extreme violence (especially with regard to the scenes that show animal abuse) and excessive length (153 minutes).

It was the film's narrative structure that attracted the attention of most critics, provoking, as well, many comparisons with Quentin Tarantino's *Pulp Fiction* (1994).[8] But as Paul Julian Smith also observes, the film has other, more immediate antecedents, closer to home. The most obvious of them is Jorge Fons' *Callejón de los milagros* (1995), also a story about Mexico City which this time draws very explicitly on life in the heart of the Centro Histórico, and which likewise has three intertwining story lines. In this case, differently from *Amores perros*, the corresponding sets of characters do know and relate to each.

In *Amores perros* the episode that links the three stories is a car accident, a most tangible and material collision. It is caused by Octavio and Jorge, who are being chased by the friends of Jarocho, a young gangster who had just been stabbed by Octavio. As the pursuers' truck gets closer and closer to their car, Octavio speeds up and dices with all the usual perils of cinematic car chases. First, they nearly hit a yellow school bus, and then they swerve right, cutting dangerously across in front of another vehicle, apparently throwing off their pursuers. Their feelings of relief are cut short, however, by the image of a truck reappearing in their rear mirror. At Jorge's instigation ('step on it, step on it'), Octavio speeds up, runs a red light and hits the side of a white Nissan. The driver of this white car is Valeria (Goya Toledo), a supermodel who has just nipped out to buy a bottle of wine to celebrate the fact that Daniel (Alvaro Guerrero), her married lover, is leaving his wife and is about to move into a new apartment together with her. The accident is observed by many people who happen to be at the scene, and some of them try to help drivers who are trapped in their cars. Among those observing the scene is Chivo (Emilio Echevarría), a street bum who was pushing a cart along, with a trio of dogs for company. The three stories in the film follow the lives of these sets of characters.

In the case of Octavio, the story tells of the events that led him to cause the accident: his involvement with his brother's abused wife Susana (Vanessa Bauche), and the fact that in order to get hold of the money he needed to run away with her, he entered his and his brother's dog Cofi in dog fights. Cofi repeatedly killed the dogs brought to the fights by the gangster Jarocho, who then shot him in revenge. Octavio stabbed Jarocho in the stomach and fled with Jorge in his car, being immediately pursued by Jarocho's friends.

The story of Valeria and Daniel, the film's model couple, goes on to tell what happened to them as a consequence of the accident: how Valeria was taken back to their new flat in a wheel chair, and how her poodle Richi fell through a hole in the floorboards and disappeared. Unable to retrieve the pooch, Daniel and Valeria get into a heavy argument. Then one night Daniel comes home to find Valeria unconscious on the floor, poisoned by wounds that have healed badly. Taken to hospital, she has to have a leg amputated.

In the case of Chivo, we learn that he is a former guerrillero turned hit man who does contract assassinations for a corrupt policeman, Leonardo (José Sefami). At the scene of the accident he is about to kill Luis (Jorge Salinas) on the orders of his half-brother Gustavo (Rodrigo Murray). We also follow his attempts to learn about and contact his daughter Maru (Lourdes Echevarría), whom he had left when he became a guerrillero and who does not know he is still alive.

The opening declaration and initial statement in the film, the collision is shown on three further occasions during it. Increasing in complexity precisely because of the manner they do and do not replicate each other, these three repetitions deftly modulate the intertwining of the three story lines. The first repetition comes after the sequence involving Octavio and Susana, and in it we see Octavio actually stabbing Jarocho after the latter had shot Cofi. It is a shorter version of the first car-chase, repeating some of the shots of the first scene, not least the view of Torre Latinoamericana, still quivering, as if to insinuate the terrible quaking of the earth that the capital is periodically subject to.[9] Fostering this possibility in deep grinding noises, the soundtrack in this case makes the scene much tenser, thriller-like. Loud music is inter-cut with silence and punctuated by a loud machine-like noise, like grinding clockwork that indicates to the viewer that something dire – the collision – is about to happen in a very short time. The machine noise is also, of course, a mechanical version of a pounding heart-beat of sorts, stressing, as Paul Julian Smith puts it, our identification with Octavio and his state of mind (2003: 63).

The second repetition of the collision occurs as we shadow the supermodel Valeria, as she leaves her lover for a moment in their new flat in order to go and get a bottle of wine, taking her dog Richi with her in the car. As she approaches the accident scene, Richi barks at Chivo's dogs (a droll message in the dog register common to all three stories) as Valeria draws up to a traffic light. As the traffic light turns green, she starts moving and we see a shot of her white car from behind, and Octavio's car hurtling towards hers and crashing full on into its side.

The third and final repetition of the collision is shown from inside the glass wall in the corner restaurant where Luis and his lover have just sat down at a table. They are being observed from the outside by Chivo, who is preparing to shoot Luis, fulfilling his contract. This time around, we are shown how Octavio's pursuers flee from the scene and how the passers-by, including Chivo, try to rescue the victims of the accident. Intimating less than Samaritan motives in trying to help, Chivo also steals Octavio's money, and takes the wounded Cofi away with him.

Shown from different perspectives and providing a link between all the main characters, the accident is clearly the culminating moment in the film. If, as González Iñárritu proclaimed, *Amores perros* is a film about Mexico City, we should then look hard at this culminating moment, as a particu-larly significant representation of that megalopolitan city. *Amores perros*, comprising the words 'loves' and 'dogs', has been accused of playing with easy metaphors (San Filippo). The one readily available to most viewers has been the dog, variously champion fighter, pooch, and vagrant in the film. In Mexico, ancient forebears were depicted in the roles of Chichimec mastiff, companion on the primeval journey beyond death, hairless meal. Of these, the conquistador invaders strongly promoted the mastiff, in their penchant for *aperreamiento*, the 'dogging' in which the other party, in chains and always the loser, was a human representative of the local population (in native script, one sixteenth-century codex records a spectacular example instigated by Malintzin and Cortés in Coyacan). In turn, the conquistador preference has been seamlessly absorbed nowadays into the binary logic of dog-eat-dog capitalism, of which modern life in Mexico City is often claimed to be a fine illustration. In these terms, we could look at the collision, set so unambiguously in relation to Mexico City, simply as a moment of social breakdown, a moment in which society, super-urbanized, canine and cynical, stopped functioning and interrupted its regular flow. It compounds, screeching to a violent halt, as if exemplifying Newtonian principles of mobility and inertia.

Yet as a collision, the accident can be more usefully seen to concentrate and compress into a single space at a particular moment human lives and loves tied each in its own way to respective dogs. What causes a car accident like Octavio's is a split-second mistake or misjudgement, which ensures that two cars occupy exactly the same space at the same instant, prompting chain reactions in those lives. In the case of *Amores perros*, this reading is underpinned by the fact that Octavio and Jorge had managed to avoid colliding twice before crashing into Valeria's car.

The first collision they avoided was when Octavio's car violently braked a few centimetres away from hitting the yellow bus. We first hear the bus's horn and see Octavio closing his eyes as if about to hit something, a shot that is then cut to a view of his foot stepping on the brakes, cut again to the a shot of his car as it is about to hit the bus, cut again to a shot showing the inside of the car and Cofi falling off the back seat ('pinche perro'), cut to a view of the bus crossing in front of car, and finally the car almost hitting it. The perfect editing of shots from perspectives that differ while holding close to that of the main character leads to our relief when 'we', that is, he and Jorge, narrowly avoid running into the bus. The second collision they avoid, shown in far less detail, happens when Octavio having cut across the path of another bus, slips into a side street where they think they have eluded their pursuers. When the accident finally happens and they run into Valeria's white car, we realize that the luck that ensured that they avoided hitting a bus in the two prior events is reversed, that is, that now a split-second coincidence brings Valeria and Richi, on the one hand, and Octavio, Jorge, and Cofi, on the other, to share exactly the same space or spot.

This concentration on the moment, on the time fragment that brings together lives that do not usually come together is what defines the relation between the accident and the city. In this sense the film is further comparable with two other films, more or less contemporaneous to it, which also portray major life changes concentrated in a split second difference: Tom Twykwer's *Lola Rennt* (1998; Run Lola Run), and the lighter romantic comedy *Sliding Doors* (1998, by the director Peter Howitt). In both films we are again presented with versions of what might have happened, in a capital city (Berlin, London), if split-second accidents or incidents had or had not happened. Both films are, like *Amores perros*, inherently urban, not because accidents happen only in cities – which of course is not the case – but because these are accidents and incidents related to a concentration of many people in a particular space. After all, the greater the number of individuals involved, the higher the chance of someone's destiny being changed in a split second. In the case of *Amores perros* we are not given alternative story lines, versions of what could have happened had the accident not occurred. But the alternatives are nevertheless insinuated and longed for throughout the film – in the way that Daniel sadly looks at Valeria as she leaves the flat before the accident (as if he is having a premonition that something bad could happen). Or in Octavio's reproachful questioning of Susana at Ramiro's funeral – asking her why she had left and cheated him (that is, taken all his money), the deed that put him in the desperate state of mind in his last encounter with Jarocho, and so on.

That three successful films produced within a few years should choose to portray urban life focusing on the concept of the accident/incident that in a split second completely changes the lives of the main characters is perhaps indicative of current perceptions of how time and space may be fragmented and yet over-determined in the contemporary city. To quote Virilio once again:

> A society that rashly privileges the present – **real time** – to the detriment of both the past and the future, also privileges the **accident**.
>
>   Since, at every moment and most often unexpectedly, everything happens, a civilization that sets immediacy, ubiquity and instantaneity to work brings accidents and catastrophes on to the scene.
>
> (1991: 56. Author's emphases)

Once again, as in Bruno Zeni's work, the timelessness of technology in the contemporary cities brings a space–time concentration focused on the moment: the moment becomes the temporal definition of life.

In the case of a megalopolis like Mexico City, the space–time concentration caused by the accident brings together not only different individuals but the social classes they respectively belong to. Octavio, Jorge, and Cofi (as well as Ramiro, Susana, and Octavio and Ramiro's mother), all come from poor parts of Mexico City, though, interestingly enough, not the very poorest slums. Valeria, Daniel, and Richi are rich, though again not from the moneyed classes at the very top of the pile. Above all, this pair belongs with the very few who have gained absolute visibility in the city through their connection with the media. Daniel is the editor of a fashionable magazine, while Valeria is a super-model whose photograph literally covers the city as part of an advertising campaign for the perfume Enchant. Static counterpart of the electronic *telão*, printed on a sheet of huge proportions, her image collapses, flops and folds dismally later in the film in the eyes of Chivo, on being repositioned by workers. For his part, Chivo apparently belongs to the extreme lower classes: we first see him as a street bum pushing his cart at the roadside, allotted not one dog but three mongrel vagrants. Our view of him in this scene is often interrupted by the passing of fast moving cars that occlude the lenses, since the hand-held camera is located on the opposite side of the avenue. In other words, Chivo first appears in the film as integral to the street, as it were, while behind him there stretches the panorama of a wretched neighbourhood. He apparently lives in a basement squat in what looks like an abandoned house or warehouse, using appliances and objects he collects from discarded rubbish. We later find out, however, that he is socially a more complex character: not only has he saved for his daughter large sums of money by being hired a killer, but he also has a past as an intellectual who left behind his life as a college professor in order to become a *guerrillero*. Captured and imprisoned for years, he finds his principles and aims in life reconstructed. He comes to play out, in precisely this dog-like or cynical sense, the role of a class-less intellectual turned lumpenproletarian gangster.

The accident is the catalyst that brings together these parts of Mexico City which normally exist in ignorance of each other. And since the accident is also what structurally holds the film together, we can say that the film can be read (as indeed it has been read) as a rendering of the social fragmentation of Mexico City. On the other hand, the accident also makes evident the fact that these parts of the city that are strange to each other nevertheless share the same space, and can all of a sudden affect each other. As Paul Julian Smith puts it: 'The pivotal car crash shows that, however segregated social classes may be through urban zoning and income differential, they cannot be hermetically insulated from one another: people will touch each other, often with fatal results' (2003: 57). Paraphrasing Zeni, perhaps we can say that if these different social classes have the possibility of affecting each other's lives in one particular moment, they do so latently all the time.

This theme becomes even clearer in González Iñárritu's latest film, *Babel* (2006), which projects on to the global stage much of what is concentrated in Mexico City in *Amores perros*. The close parallels between the two films correspond to the fact that they share not just the same director but the same script writer, Guillermo Arriaga. In extending into space whole continents apart – Africa and Asia as well as America – this sequel nonetheless remains true to the formula of the accident that transforms people's lives, and it does so adhering still to the model of three main story lines.

This time the accident is caused by two Moroccan boys who are trying out a gun they have been entrusted with in order to protect the family's herd of goats. They fire a bullet that beyond aim or intention on their part ends up passing through the window of a tourist bus and killing a female US citizen. The boys' father had recently bought the gun from a friend who in turn had received it no less as a gift from a Japanese hunter, years before. Investigation into the accident leads the Japanese police to contact the hunter's deaf-mute daughter in Tokyo. She is going through a teenage sexual crisis and is deeply affected by her encounter with the male investigator. The third story actually splits into two and involves first the American tourist and her husband, and, second, the Mexican-American nanny who is looking after their children in San Diego. As a result of the accident, the nanny finds herself having to take her charges with her to her son's wedding in Tijuana, smuggling them across the border, an experience that becomes a nightmare as they attempt to return that night.

The sheer spatial reach of the plot in *Babel* may help us retrospectively better to understand the function of the accident in *Amores perros*. Again, it is a moment of revelation of the fact that as humans we do share the same space, that our lives, however apparently differentiated and apart, always have the possibility of impinging on each other. Yet, it is precisely *Babel*'s spatial extension that threatens to make it far less effective as a film that exemplifies this hypothesis.

If the structure of different plots that are brought together by a single event – an accident – works well in *Amores perros*, it is because the characters in each story line effectively share the same space, the megalopolis. In the case of *Babel*, the sharing of global space becomes contrived. The likelihood of Japanese police actually tracking down someone who years ago had given a gun to a hunter somewhere in Africa seems rather remote, even in the present context of 'war on terror'. And because the plot relies so much on the moment that makes these lives somehow touch each other, the structure tends to break down as soon as we suspect and are not totally convinced that they are in fact touching in any meaningful sense.

In *Amores perros* the city itself stands as the location and structural element that brings and holds the three stories together. In fact, the film is full of indications that those characters, however distant in social terms and in the actual part of the city they live in, latently affect each other's lives and are made insistently to cross-reference. Valeria's image, for instance, occupies the city: she is part of the landscape and is an object of desire for Octavio's friend, Jorge. Her dog Richi barks at Chivo's dogs as they pass by. Earlier in the general plot, it is precisely because Chivo did not allow his dogs to be attacked by Jarocho's killer dog, at the beginning of the film, that Jarocho and his friends send it to attack Cofi, who then reveals his talents as a fighter. Chivo also crosses paths on the street with Octavio's brother Ramiro and his wife Susana, when Ramiro has his face disfigured from the beating he received from Octavio's friends. And so on. Contained in this mesh or on this stage vast as it is, even unnamed characters acquire the power to change the characters' destinies: Chivo's second attempt to kill Luis is frustrated because he is approached by two street children begging for money. In other words, the megalopolis is constantly being depicted as a space of social and personal fragmentation whose inhabitants, nevertheless, have an ongoing effect on each other.

In *Amores perros*, as in *Babel*, metaphor may verge on the absurd, though not in direct connection with the catalyst accident. The second story line, often described by critics as the most claustrophobic of the three, has as a central motif the fact that Richi, Valeria's small dog falls through a hole in the floorboards. Neither Daniel nor Valeria can retrieve him, although he keeps barking and crying under the floor, and that inability ends up destroying the couple's relationship. The hole in the floorboards of the newly bought apartment is another of the film's easy metaphors: its 'depth' stands in stark contrast to the superficiality of Valeria and Daniel's lives.

A sharp pointer to this superficiality is the TV programme we see before the accident, which is being watched by Jorge before he and Octavio go off to the final dog fight. In the programme, Valeria pretends to be engaged to Andrés Salgado (Ricardo Dalmacci), a (possibly) gay friend, and they have Richi as 'their child'. The supposedly 'true-life' interview is fake (it is obviously rehearsed), and the relationship is fake as well (it is a PR job intended to divert attention away from Salgado's possible homosexuality, especially among his female fans, and from Valeria's actual relationship with Daniel, a married man). As a supermodel, Valeria epitomizes the aesthetics of publicity, the fetishization of the perfect body that has become part of the landscape in urban centres like Mexico City. The sub-floor into which Richi falls – and which he has to share with rats – is the obverse of this superficial world. As Kraniauskas has indicated,[10] the hole in the floor is a central image in the film, not only because of the place it occupies in the narrative, but because it touches on the basic opposition established by the film between the superficial world of propaganda, media, and the rich, and the underworld of violence and poverty inhabited by Octavio, Ramiro, and Cofi.

Yet this 'central motif' is also absurd. Not only is it strange that a new apartment should have been sold with such a gaping hole in the floor (and have, for that matter, such a capacious subfloor), but above all that in a place like Mexico City, Daniel would not immediately find somebody to fix the problem. This touches on one of the most conspicuous absences in the film – an absence that completely contrasts with the reality of Mexico City: the absence of servants.[11] Mexico City life, and life in Mexico in general, relies heavily on servants – house maids, chauffeurs, private nurses, specialized workers of all kinds. It is virtually unimaginable that a couple like Daniel and Valeria would not make immediate arrangements for someone to free Richi, and have the house cared for by servants, and even more unimaginable that Valeria would be left all day alone in a wheelchair, without a private nurse or at least a maid to help her. The fact that Daniel claims to be short of cash (that is the reason, he says, he had nothing on the floor) still does not justify the lack of servants, since service is extremely cheap in Mexico and almost as basic for the middle and the upper classes as food. This absence or omission is so obvious, so conspicuous, that it can hardly be thought to be an oversight. More likely, it is a purposeful erasure, but to what end?

By virtue of being so claustrophobic, the central story line necessarily tends towards the theatrical and by that token can the more easily fit its allegorical function within the film, i.e., to highlight the contrast between the superficiality of the fashion world of the rich, and the deeper 'underworld' of the poor. It also underscores the social fragmentation of Mexico City: by erasing the servants, the film also erases a source of mediation, even if an exploitative one, a time-honoured intercourse and obligatory contact between the classes. It is in that sense, too, that the absence of public transport, noticed by Paul Julian Smith (2003: 56), should be read. What is left, then, is a cruder separation between the rich and famous and the poor, invisible, and hidden.

Geoffrey Kantaris correctly points to invisibility as being Chivo's best asset and most successful tool as a paid killer (2003: 188). Because he appears to be homeless, he is allowed, so to speak, to inhabit the streets, to sit in places where anybody else would arouse suspicion, even to shoot someone and not be noticed. Chivo is ubiquitous, he is part of the landscape, and in this intriguingly resembles Valeria, who would otherwise seem to be his female opposite. For Valeria is ultra-visible, the photograph of her body occupies the whole side of an apartment building, and is broadcast throughout the city in order to be noticed. So the two are both the same in being ubiquitous and opposite in visibility, a complex pairing enabled in part by the facts of life in the megalopolis. In the film, their lives are made contiguous and brought together, as it were, only by the accident. If the accident is a metaphor for rupture, for a break in the normality of the city, it is also its opposite: a moment of collision in which different parts of the city (or at least the audience that observes them) are allowed to see that they impinge on each other, that they effectively share the same space.

The importance of the accident is highlighted by the large role that cars play throughout the film. Reviewers of *Amores perros* often comment on the film's masterful soundtrack, attributed to González Iñárritu's past as a DJ, and to the talents of its sound designer, Martín Hernández (Smith 2003: 62). It is then important to notice that in a film in which sounds are very carefully used and chosen, the noise of passing cars is the most frequent sound, besides, of course, human voice. In fact, it is the first sound we hear when the film starts, when all we can see is a blurred image of the white lane divider painted on the road surface. Our first sight of Chivo, too, is strongly permeated by the sound and sight of passing cars. Many of the conversations in the film have passing cars as the main background noise, even some of the ones that happen inside apartments and houses. In fact, the impression one has after watching *Amores perros* is that the sound of moving cars is always there: it is *the sound* of Mexico City.

The scene in which we learn of Chivo's profession as a killer is a striking example of how the sound of cars is used to indicate the sound of the city itself. His victim is sitting inside a sophisticated Japanese restaurant next to the glass wall that divides it from the street in sound though not sight. Meanwhile Chivo is pacing outside the restaurant, on the sidewalk, getting ready to shoot. The scene alternates views from within and outside the restaurant. From within we see Chivo outside, cars passing by, general city movement, and in the foreground Chivo's victim eating and talking to someone else at the table. The sound inside the restaurant is soft Japanese music. Outside we also see Chivo walking up and down, and the clients of the restaurant (including Chivo's victim) behind the glass wall. The sound outside is of cars passing by. The quick alternation between soft Japanese music and the noise of cars creates a thriller-like tension, and also, a disturbing incongruity that highlights the simultaneity of both noises: on one side, the artificial, oriental oasis of the globalized world, on the other, incessant urban movement epitomized by the sound of cars, poverty, and violence. It is, to all effects, a brilliant use of sound to indicate at once the fragmentation of the city, and on the other the simultaneity of these different realities and their latent effect on each other. After Chivo shoots, from inside the restaurant we see the shattered glass and through it Chivo still holding the gun, highlighting the moment in which these parts of Mexico City are conjoined, as it were, flow into each other.

It is not just the *sound* of cars that is present throughout the film. Many decisive scenes in the film happen inside cars. We learn about Chivo's background, for instance, through a conversation between the bent policeman Leonardo and Gustavo – a conversation that happens entirely in a car, in a long trip through city avenues. Similarly, when Chivo kidnaps Luis, we are taken through the city in the latter's Mercedes. We are made aware of Daniel's unhappy marriage in a car scene, when he is driving his wife and two daughters home. And we also see the beginning of the deterioration in Valeria and Daniel's relationship when he is driving her back from hospital and Valeria complains (in her strong peninsular Spanish accent) that he keeps on saying that things will get better. The wall in Octavio's bedroom is decorated with small stickers portraying cars, and one can see model cars spread throughout the room. The only big item Octavio buys after he earns a lot of money in dog fights is a car – the same car that causes the accident later. When his brother Ramiro comes to his room to demand equal shares in the profits made at the dog fights, Octavio is watching a programme that shows computer images of cars being smashed on top of each other. Chivo takes Gustavo's and Luis's cars to a scrap yard, and one shot of the yard shows, as on Octavio's television screen, cars piled on top of each other, being destroyed.

In 1996 Castells described global economy in terms of flows:

> Our society is constructed around flows of capital, flows of techno-logy, flows of organizational interaction, flows of images, sounds and symbols. Flows are not just one element of the social organization; they are the expression of processes dominating our economic, political and symbolic life.
>
> (1996: 412–413)

The megacity of *Amores perros* is best defined by the flow of cars: cars move back and forth in the city, define its pace, its sound, its movement, and its class relationships. They are also acquired and disposed of. Curiously, there are no images of Mexico City's famous traffic jams: cars in the film are always moving, making the same insect-like buzzing noise, passing by.

Although there are images of people walking – and Chivo, above all, is often portrayed walking the city, though he also has and drives an old truck – the scenes that show people in cars easily surpass in number those of walkers. As Paul Julian Smith says, the film shows no images of public transport, nor even of the green and white taxis that in fact are to be seen everywhere in the city: '*Amores Perros* shows not the street as a site for social intercourse but the road as a conduit for anonymous traffic . . . Characters are confined to private cars, which are in the city, but not of it' (2003: 56). In other words, the remarkable predominance of cars in the film highlights, once again, the fragmentation of the city, since though these vehicles may move more freely than those in São Paulo, they seldom fail to enclose, encase and compartmentalize individual experience.

Moreover, cars and car driving suggest another 'practice of everyday life' that is complementary, if not alternative, to the one described by Michel de Certeau in his much quoted chapter 'Walking in the City'. In it, de Certeau reminds us that the ordinary practitioners of the city 'are walkers, *Wandersmänner*, whose bodies follow the thicks and thins of an urban "text" they write without being able to read it' (1984: 93). Pedestrians, according to de Certeau, 'weave places together. In that respect, pedestrian movements form one of these "real systems whose existence in fact makes up the city". They are not localized; it is rather they that spatialize' (ibid.: 97). De Certeau never includes cars in his analysis, although he does refer to the enclosed space of the train carriage as 'traveling incarceration' (ibid.: 111). As Thrift puts it:

> But, as de Certeau would have surely underlined, this system of automobility has also produced its own embodied practices of driving and 'passengering', each with their own distinctive histories often still waiting to be written. Though we should not of course forget that how the car is put together, how it works and how and where it can travel are outwith the control of the driver, yet it is still possible to write a rich phenomenology of automobility, one often filled to bursting with embodied cues and gestures which work over many communicative registers and which cannot be reduced simply to cultural codes. That is particularly the case if we are willing to travel off the path of language as the only means of framing communication . . . and understand driving (and passengering) as both profoundly embodied and sensuous experiences, though of a particular kind, which 'requires and occasions a metaphysical merger, an intertwining of the identities of the driver and car that generates a distinctive ontology in the form of a person-thing, a humanized car or, alternatively, an automobilized person.
>
> (2004: 47)

In *Amores perros*, cars are often the place from which the city is seen and experienced, and where relationships, as we have seen, unravel. Chivo often appears observing his victim-to-be and his daughter from his old pick-up truck – and both Luis and the daughter are also seen in their own separate cars. When he finally gets hold of Luis, Chivo handcuffs him to the steering wheel and makes him drive through the city, towards Chivo's house: the car, which up to then had provided Luis with mobility and status, becomes, with the kidnapping, his own prison. In many Latin American cities (Mexico City and São Paulo being certainly the case), kidnapped victims are indeed mostly taken away in their own cars, and it is the cars themselves that often reveal the victim's wealth to the kidnappers: in other words, it is because they drive expensive cars that many people end up being victims of kidnap. Cars in the film are an extension of the characters' bodies: Jarocho drives a menacing-looking pick-up truck with volutes of flame shielding the sides; Valeria drives a slim imported car with a radio that is playing a tune that is a catchy, maybe, but mindless synthesized Hispanic version of a US model; Chivo drives a wreck, Luis drives an expensive Mercedes, and so on. It is in cars, or near cars, that much of the conflict takes place – not least, as we have been emphasizing, the scene that unifies the three plots. If, following de Certeau, we try to understand the 'lines' traced by the movement of cars in the film, we will see that they point, like the rest of the film, to class difference and conflict, and to compartmentalized lives unaware of each other. Cars also shape the city, as much of the infra-structure that is emphasized in the film has to do with their circulation: wide avenues lined with concrete, multiple-lane streets, traffic that slips between the camera and the characters, streets as illuminated by public and car lights, and so on.

The fragmentary social and economic reality of Mexico City also pervades the most notable motif in the film: the dogs proclaimed in its title, their fights, already contemplated above as an age-old local metaphor and corollary for human behaviour. Kantaris describes dog-fighting in the film as displaced violence: 'Yet the dog-fighting is clearly acting as a displaced metaphor, an allegory even, for human violence in the film, and indeed dog-fetishism generally in the film takes the form of a displaced and dislocated cipher and substitute for impeded human relationships' (200–3: 186). Following from that, we can say that dogs also act as 'displaced cipher' for the profound inequalities of the megalopolis. On the one hand we have the tiny Richi, the frail, almost doll-like lap dog, professionally groomed and over-protected procreation substitute for Valeria. At the other extreme, we have the fighting dogs that Jarocho 'spends', one after the other, in his attempts to make money and to affirm himself against Octavio. Kantaris correctly defines the business of dog-fighting as 'an allegory for the systemic violence inherent within globalized capitalism' (2003: 186–7). He goes even further, saying that

> The owner of the dog-fighting business runs his firm according to strict and sound neo-liberal, market-oriented practice. 'Esta es mi empresa' [This is my business], he explains to the two new lads, 'no pago impuestos, no hay huelgas ni sindicatos, puro billete limpio' [I don't pay taxes, there are no strikes nor unions, all hard cash]. As an example of no-barriers private enterprise, of 'flexible' accumulation operating with a labour force battered into passive submission, I am sure that the IMF would no doubt approve most highly.
>
> (ibid.: 187)

The argument being proposed here would not easily stretch so far as to equate the dog-fighting business with neo-liberalism, though it certainly stands for at least one aspect of neo-liberal economy, which is its reliance on underground

economic activities, and its propulsion of a large part of the world's population (above all uneducated young males) into an underworld of jobless
illegality and violence. The differences between the owner of the dog-fighting
business and neo-liberal economy lie, first of all, in his open declaration that
he does not pay taxes. Although neo-liberalism favors tax-cuts for the rich,
it has also increased taxation in the form of sales tax, and cuts in tax breaks
for the poor. But it is above all Mauricio's reliance on 'billete limpio' (hard
cash) that must surely constitute the main difference between his business and
neo-liberal practice. Neo-liberalism is largely based on a financial order that
has very little connection with 'billete limpio'. Instead, it relies on the highly
fictional ups and downs of the stock exchange, credit, debts, etc. Mauricio's
exclusive reliance on hard cash is connected with the ever increasing importance of the so-called 'informal economy' in the current world, and above
all in its most underground arena, drug traffic, smuggling illegal weapons,
and so on, which is (and here Kantaris is right) effectively continuous with
neo-liberal trade and politics.

Like drug traffic, dog fighting is illegal and underground. But while the drug
traffic wastes the lives of thousands of young men year after year, throwing
them into a violent maelstrom whose only function is to produce vast wealth
for very few, dog-fighting does the same thing, but with dogs. Moreover,
dog-fighting in itself is a sub-product of urban violence: killer dogs are trained
to be employed as security guards against the dangers of the megalopolis.
Dog-fighting uses this security mechanism as an end in itself: dogs are
trained not to kill for security, but to kill each other in order to provide money
for their owners. What makes the film so effective is precisely what Kantaris
has called a displacement mechanism: it allows viewers to see the killers as
absolutely brutal and ruthless and, at the same time, innocent and exploited.
This could never happen if the film dealt with poor young men. The core
metaphor of the film, the dog-fight by necessity totally devoid of 'love',
stands for the 'dog-eat-dog' life in one of the largest cities in the world, where
it can be extended to 'dog is exploited by humans to eat dog in order to
produce money'.

The film's use of urban architecture also emphasizes the fragmentation and
isolation of life in the megalopolis. As we have seen, *Amores perros* hardly focuses
on any of the most characteristic landmarks of the city. Yet, it is a highly
architectural film. House exteriors and interiors are used to denote social class,
intellectual formation and taste, religious affiliations and so on. Exteriors
are particularly revealing of modes of life in contemporary Latin American
cities. In this regard, frames of certain exteriors, above all those of three houses
or abodes, are especially eloquent in what they tell us about the characters
associated with them: Susana, Chivo's daughter, and Chivo himself.

Early on in the film, immediately after the first dog fight, we are presented
with a standing frontal shot of a house. Even though it is taken from the
opposite side of the street (and eventually interrupted by the passing of cars)
it is a flat, almost photographic shot, both because of the standing camera

and because the house is flush with the sidewalk and offers very little perspective to the viewer. We then see a schoolgirl approach the house, open the front door, and a dog escape from the door. The camera continues without moving while the girl calls the dog, leaves the reach of the camera to go look for it, and comes back and enters the house saying 'pinche perro' (bloody dog). Later we find out that the girl is Susana and the house is where Octavio's family live. It is a reasonably long shot for a fixed frame, and it seems particularly so because it comes after two initial scenes filmed mostly with hand-held cameras (the car chase and a dog fight). The house is a poor (though not extremely poor) two-storey construction covered with green tiles outside and no tiles on top. Its generally dirty front indicates that it is located in an urban, polluted place. Above all, the fact it is a terrace house points to the possibility that there are other houses similar to it to either side, though we never see them. There is a strong sense that this is a house like any other one, a house anywhere in the poor parts of the city.

Along the same lines, we are presented with several shots of Chivo's daughter's house from the outside. The architecture is once again unremarkable, except as a clear demonstration of the phenomenon described by Teresa Pires Caldeira in *City of Walls* with regard to São Paulo – that is, the presence of tall gates and bars that completely obscure the façade. In fact, when we see Chivo's daughter emerge from the house, especially in the second shot, we cannot but be reminded of a prison. The relationship between the inside and the outside in this upper middle-class environment that Chivo had forsaken is marked by the architecture of fear, as Mike Davis describes it in his early studies of the 'fortress city' of Los Angeles (2006: 223–4).

Yet, that architecture proves strangely similar to the place where Chivo himself lives, the third architectural sample. Entered through a small door set into a high solid gate, his home, like his own daughter's house, is reminiscent of a prison. Its function, in this case, is not only to protect the interior from the dangerous outside, but also to guard the mysteries cloistered in the underworld that Chivo represents. The three façades that shelter Susana, Chivo's daughter and Chivo emphasize the isolationist architecture of a city where, as Paul Julian Smith observed (see above) we are given almost no examples of conviviality, and certainly no examples of neighbours talking, greeting each other, or in dialogue. The altercation between Octavio and Jarocho, far from casual and further from friendly, would be the exception that proves the rule.

In the neighbourhood where Octavio and his family live, we are invited to consider a further example of architecture, now, however, a home that downstairs doubles as its notional opposite as a place of conflict: the dog-fight arena. The first view we have of it is internal, and focuses on the patio where the fights take place, defined by a narrow blood-spattered wall. In a later scene, the owner, Mauricio, takes Octavio and Jorge to the rooftop terrace (*azotea*) to negotiate the deal concerning Cofi. From there we can see a confusing mesh of unplanned, unfinished and unpainted constructions and patios covered with corrugated plastic, out of which there energes what look like the towers of a colonial church. This view of modest houses that multiply as far as the eye can see is as much a 'typical' view of Mexico City as any of its landmark monuments. As Carlos Monsiváis declared in *Los rituales del caos*: 'beside the Teotihuacan pyramids, the baroque altars and the areas of elegant Mexico, the popular city projects the most typical image – i.e. the brutally massified image – of the coming century' (1995).

Similarly, many of the scenes, including the accident, happen in posher commercial and residential areas of the city that, although much more affluent than the neighbourhood where Octavio lives, are not so exclusive as very wealthy neighbourhoods such as the Lomas. They could be anywhere in the fashionable restaurant areas of the city: colonias like Roma, Condesa,[12] and so on. Once again the film plays with both a lack of clear landmarks and the choice of urban landscapes that are very typical of Mexico City.

This concentration on a generalist view of Mexico City reinforces the film's emphasis on fragmentation: characters like Octavio, Ramiro, Susana, Chivo, and Jarocho resemble millions of others in the city in living out lives that as a norm are submerged in megalopolitan anomie. The accident brings them out, changes them, makes us (if not them) aware that lives lived in completely different quarters of the city can impinge on each other. But even the accident, which happens at a common corner of a city dominated by cars, is just one among many, one more unremarkable event, a moment in the life of the megalopolis.

The film does not end without duly tying up some threads of the plot, helping us to complete the crossword puzzle. Chivo leaves the most unfraternal half-brothers Luis and Gonzalo with their hands tied, facing each other at opposite ends of a room, a gun between them. He holds on to Cofi, and none escapes the implacable dog-eat-dog logic of the place. In rounding off the action, the final scene of the film enhances its dimensions, above all in time. Chivo is shown entering a scrapyard in the company of his newly acquired Cofi, who has killed his other dogs and is christened 'Blackie' (the devil's canine colour). He has come to sell Gustavo's SUV and having done so he sets off, walking eastwards towards what appears to be early morning light in the sky above the volcanic rim of the megalopolis. As they lengthen, the lingering shots of him from behind intimate a lengthening of time, invite access to the deeper history of that place.

The terrain Chivo crosses is deserted, dried up and cracked as if once covered with water. This reminds us that the island city of Tenochtitlan was built in a lake basin which the European invaders, glorious as they recognized it to be, lost little time in attempting to drain. The route he chooses neatly reverses that of Cortés and his men when they arrived from the east, passing between the sentinel volcanoes. As the story ends, history begins to resonate in the film, amplifying its registers. Walking away from the city, cynic Chivo leaves behind its synchronic, dog-eat-dog temporality, allowing glimpses of the far far longer experience proper to this megalopolis.

3 Flânerie

POST-APOCALYPTIC

Chivo is the only persistent walker in *Amores perros*, his habitual 'invisibility' allowing him to play the role of an observer in the crowd. When we first see him in action, about to make the first contract killing featured in the film, he is pacing up and down in front of the restaurant, separated from his victim by a glass wall (see p. 49). Seen from inside the restaurant, he looks just like a street-bum staring hungrily at luckier people eating away to their hearts' content. After shooting his victim, Chivo runs away and nobody tries to catch him. In fact, nobody even looks at him: he slips into the crowd, a proverbial fish into water, like the *guerrillero* we learn he once was. Similarly, on the two occasions when he breaks into his daughter's house, he is able to do so in broad daylight, without catching anybody's attention.

Chivo is invisible not only because hunger and homelessness are common features in the space of Mexico City, but also because, although common, they are bothersome and disturbing for the middle-class passers-by, who tend to look away, avert their gaze. Throughout the film there are many shots in which Chivo is seen observing his victims, or his daughter, in busy streets, with cars and people passing by, without ever being noticed, the observer unseen and unobserved. He tails Luis closely several times, as the latter walks with his lover to a restaurant or to a motel, or as he is affectionately embraced by his half-brother (and Chivo's contractor), and so on. The eyes that observe the crowd without calling any attention to themselves, identifying perhaps with particular elements in it, are like those of the *flâneur*. Yet they simultaneously invert the sight lines characteristic of Baudelaire's *flânerie*. Instead of channeling the gaze of the poet who compassionately observes the poor and tries to feel what they feel, Chivo's eyes look back. They are, quite literally, 'the eyes of the poor' (as in the title Baudelaire's poem), which, justifying the urban paranoia of our time, stalk and threaten the rich. Moreover, we learn later in the film that Chivo is not just a street bum: he is also a former college professor turned *guerrillero*. He is educated, articulate, well informed: an enlightened observer of crowds. And he is a cruel killer, who moved from the political violence of the 1960s – violence with an ideological end (as he says to his daughter in the recorded message he leaves for her: he wanted to change the world) – to the senseless violence of the contemporary megalopolis, where money is the only thing that counts. If Berlin was described by Benjamin as a city that transformed the philosopher-*flâneur* into a werewolf, what to say of the contemporary megalopolis, where the *flâneur* has become both destitute and a paid killer?

Issues of class and *flânerie* also inform Armando Ramírez's *crónicas*, notably 'Muerte anónima: el Rasguños' and 'El amigo de Catarino Vélez regresa al lugar de los hechos . . .', both from the collection *Bye bye Tenochtitlan* (1991 – a remote echo of Isherwood's *Goodbye to Berlin*?). As we shall see in Chapter 4, Ramírez is known for his *crónicas* and novels of and about the Tepito neighbourhood, which develop a geography of the city based on a strong division between the places where the poor live (Centro Histórico and the areas immediately to north and north-east) on the one hand, and what he calls *el Sur* (the South), coded as the parts of the city inhabited by the intel-lectuals and the rich. Tepito features large later on (p. 82); here the focus is on how *walking* and the idea of *flânerie* are inscribed by Ramírez onto its keenly dissected social geography.

In general terms, *Bye bye Tenochtitlan* is the most nostalgic of Ramírez's books. It is composed of three parts: 'Las noches de México' – a series of *crónicas* that explores the subject of night life in the city, above all, cabarets and prostitution; 'Personajes de la ciudad', which, as the title suggests, deals with certain characters in and around the centre, and 'Lugares de la ciudad', which concentrates on life in a few chosen locations in the city. In a short 'Prólogo' the author defines the book's nostalgic tone, and the problematic location of the narrating voice:

> The old city, the old plan, the Centre, the beautiful Anáhuac, la Tenoch . . . It is not what it used to be . . . it is . . . is . . . other, the other, the same and not the same! How do you see it from there?
>
> The other one is the other's one. Not yours, you are me, I am you, the other.
>
> Your city belongs to your time, your mind, the one you live and will remember, the one you recognize yourself in. The other one? It is the other's one.
>
> . . . .
>
> Am I or do I resemble myself? Do you see me or do you guess me?
>
> (1991: 90)

Defined through the author's nostalgia, the city is at the same time 'one' and 'the other', that is, the city one remembers or has learned about – the city of memory – and the city one sees and lives in the present – the city of the walker, the *flâneur*. The narrator's divided self is thus a reference to mem-ory and history, to the city's transitions and changes. It is, at one and the same time, a reminder of the narrator's simultaneous identification with his characters and his detachment from them – a process that is not dissimilar, for that matter, to Baudelaire's *flânerie*: 'For Baudelaire, *flânerie* was an indispensable modality for reading urban modernity, it allowed a form of spectatorship that was at the same moment both detached and immersed in the rhythms of the crowd' (Highmore 2005: 40).

In 'El amigo de Catarino Vélez regresa al lugar de los hechos...' (Catalino Vélez's friend returns to the scene of the crime . . .), for instance, the narrator talks to his character Pancho Betancourt, explaining episodes of Pancho's own life:

> Anyway you look at it, Panchito, you can't lose, you had been at this exact spot where you found yourself, but that was the spark that enlivened your life. Twelve years ago? Why am I telling you this, when it's up to you to count up the years, of what you call your bad luck.
>
> (1991: 105)

Throughout the *crónica*, the narrator simultaneously tells his readers what his character feels and experiences, sympathizes with him, questions his character's ability to understand his own experiences, and doubts (as in the example above) his own capacity to speak for his character. While Pancho walks the city, the narrator observes him, scrutinizes him, explains him to himself, but also is ironic about his own position. The technique highlights the class/educational differences between the narrator and Pancho by presenting a double look at *flânerie*: on the one hand, it focuses on the writer who describes city characters, and on the other it accompanies, with sympathy, the experiences of a poor man who tries to walk in the city. Detached, as it were, from the crowd, the narrator can understand and explain experiences that his character supposedly cannot. He emulates, in that sense, the enlightened poet/*flâneur*: as Keith Tester observed, in Baudelaire, the poet is the only individual in the crowd that knows that he is alone (1994: 10). But, as narrator, Ramírez also identifies with his character, professes to understand what he feels, presents his character's own views of his experiences in the city. The man we see walking in the city is not the poet, but Pancho. The poet is an implied walker, as it were, a man whose experience of the city can be deduced from what he has to say about, and through, his urban characters.

Differently from Baudelaire's 'gentleman of leisure' (Benjamin 1999b), Pancho can only allow himself to walk because he decides to take a day off work. A man in his thirties who sees himself as old because he has been submitted to an oppressive routine (long working hours in the metal industry, early marriage, four children, and no money or time for pleasure), Pancho is not destined to be a *flâneur*. He does not even know the streets well, the first of the sample given in the text meaning aptly enough 'lost boy': 'Calle Niño Perdido, Santa María La Redonda, San Juan de Letrán, these street names would get all mixed up in his head from the time he was a child, he could never tell San Juan Letrán from the others.' His desire to take the day off in order to walk is related to his memories as a young man, when he used to stroll San Juan de Letrán Avenue with his friends, looking for his first sexual experiences. Since then, San Juan de Letrán has become Eje Vial – a wide avenue that literally cuts the north of Centro Histórico in two (just as railway lines cut up Dickens's London), and has incorporated the other two streets from his childhood: Niño Perdido and Santa María la Redonda. The change in the streets' names and layout stands for other changes in the city: it is now completely organized around the movements of cars, hostile therefore to poor walkers like Pancho.

But it is when Pancho approaches Torre Latinoamericana that we see how dangerous public urban spaces can be for walkers like him. The same building that appears in the opening scene of *Amores perros* (see Chapter 2), Torre Latinoamericana is definitive as marker in Mexico City's skyline, just west of the centre. It was the tallest building in the city between 1956 (when it was built) and 1984. During these years it was *the* emblematic image of modern urban Mexico. Next to it stands the Procuraduría General de la República, the national police headquarters, of ominous repute. As Panchito looks at the latter, the narrator, making use of free-indirect speech (in other words, moving the narrative between the dialogic and the monologic modes) warns him:

And who are they, those people who grab Pancho as soon as they see him. More than twice they have forced him into their cars, exactly when he was leaving the factory, wages in hand. Come on, Panchito, don't let your knees shake, you know anyway you are prisoner number one hundred, and as you stroll by on your day off work on a Monday you're not going to get away, you're going to be grabbed, you've been marked by God's finger, you've got no influential parents, no relatives who count and you don't live in the South of the City.

(1991: 117)

For Panchito, a common citizen who has no influential parents and who does not live 'on the South side of the city', the mere act of strolling, of walking the city outside the limits of his neighbourhood, is an invitation for police brutality. The narrator (or Panchito's own voice, if we take the use of free-indirect speech to be an indication of monologue) tells Panchito to get lost in the crowd

It is best for you to get in the middle of the tangle of people who at this moment are going to walk past the Procuraduría, so that, mingling in the crowd, the police officers won't recognize you and know you are a worker, the kind of worker who lives at the foot of the hill in Villa Guadalupe Hill [that is, in Tepito – see Chapter 4].

(ibid.: 117)

For poor people like Panchito, getting lost in the multitude is not a way of feeling – as it is for Baudelaire's *flâneur* – that one is part of the city. On the contrary, it is a way of disguising the fact that he does not feel entitled to the modern city, that he does not belong in it, he is not a 'citizen':[1]

> Get in the middle of the masses, don't allow yourself to be seen, let the multitude shelter you, get into anonymity, so that they don't know you are ignorant, or that you have no money, or that your relatives will go crying and pleading, wringing their hands at the prison gates, rather than ringing someone important.
>
> (ibid.: 117)

Hence the title of the *crónica*, a reference to Panchito's childhood friend Catarino Vélez, who years before, and in the same part of the city where Panchito is now walking, had 'disappeared inside a car with no number plate just because he lived in Ciudad Nezahualcóyotl and carried a bag with his mason's tools' (ibid.: 118). The only times that Catarino Vélez is mentioned in the *crónica*, after the title, are through Pancho's memory. He does not play any actual role in the text, yet it is his name that gives the piece its title, as a reminder of what can happen to the poor who dare to walk fearless in the modern parts of the city. The police put a generalized liberal prejudice into practice yet also recall ancient custom in these very streets, where the poor would shrink into the shadows at the approach of the Mexica knights (as Clendinnen (1991) reminds us).

'Muerte anónima: el Rasguños' makes the same point even more clearly. Narrated in the third person, this short and delicate *crónica* tells the story of Rasguños, a street child who was killed by the police. Rasguños's identity is defined, from the very beginning of the *crónica*, by the megalopolitan street:

> He grew up in the street. Sometimes, the street appears to be clandestine. He, fourteen years old, died legally, but for others it was a clandestine death. And not because he was a *guerrillero*, someone who actively tries to bring a government down. Neither was he politically informed or a slogan-painting student with subversive tendencies. He was another kid who wandered the streets crammed with almost twenty million human beings.
>
> (1991: 77)

By calling the street 'clandestine' (*clandestina*), the narrator questions its definition as a public space. The term 'clandestine' has very clear resonances in recent Latin American history, which are invoked by the narrator. The *guerrilla* movement of the 1960s and 1970s was often termed 'clandestino' and those who joined the *guerrilla* or who were persecuted by the Southern Cone dictators (in Brazil, Argentina, Chile, Uruguay, and Paraguay) and would go into hiding, that is, they would become 'clandestinos'. The act of walking on the street, of openly being on it, is the opposite of being clandestine. Yet the street effectively becomes clandestine for those who are not allowed to be in it whenever they want – paradoxically, perhaps, the people who are closest to it: the poor.

Rasguños grew up on the street or, as the narrator says, 'on the margin, in the shadows, on the dying side of life' (ibid.: 79). He was used to walking on the streets and preferred to do so late at night, at three or four in the morning, when he would be spotted 'viendo viendo a la nada' (looking looking at nothing). The gaze that apparently looks at nothing on city streets reminds us again of the *flâneur* and his gaze. But what Rasguños saw is first described as the reverse of busy city streets: 'he liked the city like this: cold, quiet, moribund, no policemen around, no people' (ibid.: 82). It is also described as movement, but of another almost supernatural order: 'He would walk and see the dead and the ill, the wounded and the drunk, the policemen and the thieves, it was like being in the secret routes of life's days' (ibid.: 82). The expression 'secret routes' also indicates clandestine nature: the night life that Rasguños observes is the clandestine life of the city. The theme is in fact dear to Baudelaire's *flânerie*: the poet/*flâneur* was after all often seen straying into the worst parts of the city, observing prostitutes, beggars and thieves. What is different from Baudelaire are the eyes that observe the city, now the eyes of those formerly observed, the destitute, marginal eyes of the street child. Likewise the extremes of violence and urban indifference that those eyes observe in the megalopolis: one day Rasguños saw policemen 'dragging pieces of bodies into the gutter' (ibid.: 82). Another day, 'a dog was eating bits of rotten meat from a rubbish bin; a man arrived and cut the dog into pieces with his axe'. On yet another night

> [A]*mariachi* lay dead in the middle of the street, and the fast cars ran over over him like a bump on the surface, he counted thirty-one cars before the police arrived and found the body. He was sitting at the edge of the gutter, watching how the traffic slowed down, how people looked and moved away.
>
> (ibid.: 82)

The main difference, however, is that Rasguños is not always entitled to observe the city: every time he sees a policeman, he has to move away or attempt to hide. At the night of his death, Rasguños saw the police arresting a cab driver, and pretended not to see it. He then heard the sounds of yet more police cars:

> He felt above all the noise of violence, the red and blue lights, sirens piercing the silence, he felt his body tense, he saw the men in suits and pistols shattering the pools of darkness, they arrested people, they beat them and put them in cars and vans; Rasguños told himself: another raid.
>
> (ibid.: 82)

The police raid turns the street into a space of repression: anybody who happens to be on it can be detained. During a police raid, not just thieves and outlaws, but passers-by, walkers, workers, and *flâneurs* – all lose their right to be on the street. The police action is described as irrational, as indiscriminate beating and abuse, and Rasguños, who had seen his father (a thief) being almost killed by the police, tried to run away. He was shot from behind: the police claimed he was told to stop and did not obey: hence the 'legality' attributed by the narrator, ironically, to his death. Since Rasguños's only relatives were his siblings, there were no complaints: 'Nobody said anything, it is secret, the street was empty, his siblings cry, but they know, but nobody knows, death is dark' (ibid.: 83). Rasguños's death is ignored by all: 'Neither the newspapers, nor the television, nor the radio, nor the people, nor the police knew anything, he did not even have a birth certificate. In life, Rasguños was like a clandestine being' (ibid.: 83). While at the beginning of the *crónica* the street is described as being, at times, clandestine, now it is Rasguños that merits the adjective. He bears a metonymic relationship to the street. Like the *flâneur*, he is part of the street, but unlike the *flâneur*, he does not merit the automatic right to walk on it when he pleases.

The same theme appears in a short film set in São Paulo, Mirella Martinelli's *Opressão* (1993). Tatiana (Gabriela Cardoso), the protagonist of the film is doubly oppressed: because she is a woman, and because she is poor. Many of the scenes in the 13-minute film show Tatiana walking on the streets of São Paulo late at night or early morning. She cannot be properly called a *flâneur*: she has no time for it, she walks through the city mostly trying to get from one place to another. The mother of a small baby, she starts the film in her flat waiting for the baby's father to arrive and take over childcare. Luís (Márcio Trinchinatto), the father, arrives, only to tell her that he cannot look after the baby that evening. A performer of contemporary dance who works at night, Tatiana then has to take the baby to her mother's. On the way to her mother's house we see Tatiana walking and waiting for a bus underneath a concrete bridge, being observed by drug addicts. She is tense, looking around herself. The street behind her is shot diagonally, as if falling – a world in disarray. At her mother's, Tatiana has to hear complaints about her job, the father of her child, the fact that she had a baby with a man who has no proper job. Back on the street she walks and cries, having as background peeling and decaying textures of city walls.

The night club where Tatiana works is intensely urban: a basement-like room filled with gothic or punk couples dancing to the sound of Nina Hagen or the Brazilian rock band *Os Inocentes* singing a number entitled 'Desequilíbrio' (Imbalance). Tatiana is dancing with the lead-singer of the band, a black man, when a group of neo-Nazis walks in and starts beating performers and public, concentrating their wrath on the black lead-singer, who is left unconscious. At dawn, after leaving the hospital where their friend is about to die, Tatiana and the gay owner of the night club stop on the street to smoke marijuana. A group of policemen arrives, searches them violently and takes them to the station, threatening to kill them and to rape her as they drive them away. After spending some hours in jail, Tatiana is shown again walking, and then waiting for a bus in front of a bar. A drunkard, not less a victim of oppression but armed with a gun, makes an indecent approach to her. She answers indignantly and aggressively, and is shot in the head. The last scene shows the passers-by approaching her body, filmed from below, from her point of view. Their cries of indignation and offers of help become distorted as the camera is tinted with red. Tatiana dies.

The short film has an apt title: it is hard to believe that so much oppression could fit its 13 minutes. Yet, if the concentration of so many oppressive events in one single night may seem unrealistic, it fulfils the film's allegoric purpose of looking at the theme 'oppression' in many of its forms: through gender, race, politics, police, work, motherhood, etc. All those forms of oppression are inscribed on to the film's urban space. Before arriving at the bus stop where she is finally going to die, Tatiana is seen crossing a large avenue full of cars: she has to cross diagonally through various lanes of stopped traffic, climb over a concrete barrier, negotiate her way between several further lanes of fast moving traffic before she gets to the other side. Such a long scene in a short film like *Opressão* serves as a metaphor for Tatiana's own trajectory. The oppression of the film's title stands for the oppression of the megalopolis where a young woman – herself very urban in her attire, her musical and dance tastes – tries to make a living. Though a large proportion of the film tracks Tatiana as she walks on the streets, she never does so with pleasure: we see her crying, worrying, afraid, being molested by the police and by the drunkard, and finally being killed. She is never an observer: she is the one who is constantly observed, stared at, leered at, harassed and pursued.

Feminist and non-feminist critics have long been discussing whether or not women can be *flâneurs*. Anke Gleber, for instance, argues that women 'did not have the right to look, to stare, scrutinize, or watch' (1994: 4), and although in the nineteenth and early twentieth century there were women painters and many women walking on the streets, they required

> an additional physical and psychological confidence: the initial courage to step out, face the threat of assault, or erase her latent misrecognition as a prostitute, in short, to undergo the constant encounter and annoyance of being made into and treated as an object.
>
> (ibid.: 34–5)

E. Wilson, similarly, claims that women

> along with minorities, children, the poor, are still not full citizens in the same sense that they have never been granted full and free access to the streets . . . . and they have survived and flourished in the interstices of the city, negotiating the contradictions of the city in their own particular way.
>
> (1992: 8)

Ramírez's *crónicas* and Martinelli's short film show precisely the limits of *flânerie* for those – women and the poor – who have not 'been granted full and free access to the streets'.

Such, however, is clearly not the case for Carlos Monsiváis, Mexico's megalopolitan pundit *par excellence*. His *Los rituales del caos* (1995) might be described as a contemporary exercise in quite another style of *flânerie* in contemporary Mexico City. Monsiváis is not only a *cronista* who writes regularly for Mexican newspapers, he is also one of the best-known intellectuals in Mexico. Openly gay, humorous, very critical of the authoritarianism and corruption in the Mexican government, Monsiváis is respected in both academic and non-academic circles. He is as likely to appear on television (and be recognized in taxi cabs or on the street) as on examining boards at the universities. He seems to be surrounded by a virtually unanimous recognition that no intellectual in São Paulo, or even in Brazil, is able to boast. He is particularly respected for the courage he has shown, investigating homicidal operations ultimately traceable back to the government.

As a respected intellectual, Monsiváis is obviously able to walk freely (though not necessarily invisibly) in most parts of the city. He also knows that walking is not the only – perhaps not even the most obvious – way of experiencing the contemporary megalopolis. The focus of *Los rituales del caos* is the multitude precisely as it is experienced both physically and via the media. For Baudelaire, observing the multitude was the best remedy for boredom: 'any man who can still be bored in the heart of the multitude is a blockhead! A blockhead! And I despise him!' (1986: 20). The multitude was, according to Benjamin, the 'veil through which the familiar city beckons to the *flâneur* as phantasmagoria – now a landscape, now a room' (1999a: 10). That is, for the *flâneur*, the multitude rendered the experience of being in the city, both familiar and unfamiliar.

Monsiváis cannot immerse himself in the multitude. He looks for it not in casual corners of the city, in the everyday movement of passers-by, but in what he calls 'rituals', that is, moments which inspire gatherings, pull together immense numbers of people, like a boxing match, a football game, a rock concert, a fiesta in honour of Guadalupe or similar faith-based occasions, a conference on self-improvement, though not in these *crónicas* the kind of mass political demonstration that periodically takes over the Zócalo entirely.

Besides the initial vignettes (in italics), the only text in *Los rituales del caos* that does not describe an organized event that brings together a multitude is a short humorous *crónica*, inset into the middle of the book, about the crowded metro. In it, he jokingly defines Mexico City metro as a place where two objects can occupy the same space at the same time: 'In the Metro the molecular structure loses its universal empire, anatomies melt together as if they were spiritual essences, and transcorporeal combinations impose themselves' 1995: (112). In her commentary on this *crónica*, Jean Franco observes that

One of Monsiváis's most interesting conclusions is that individualism based on the integrity of the body is no longer possible in a mass society. The Mexican subway is so crowded that the limits of the body that had once defined the individual disappear in the mass so that even the sexual pleasure of bodily contact is diffused.

(2002: 99)

Interestingly, the *cronista* does not situate himself in the metro at the moment of writing (as he does, for instance, in the case of the boxing match, the rock concert and other *crónicas*). And although the metro is defined in the *crónica* as 'the city' and as a 'staging of the meanings of the city' (Monsiváis 1995: 111), the fact is that for the most part it is used almost exclusively by the poor (students on the Ciudad Universitaria line being a partial exception).

Throughout the book we follow the *cronista* as he watches multitudes that watch sports events or concerts, or participate in collective festivities. Although he occasionally uses the term 'massas' (masses) to refer to some of those moments, it is 'multitud' that appears most frequently in his *crónicas*. His definition of the multitude is centred on the idea of spectacle. Watching a boxing match between Julio Cézar Chávez and Greg Haugen, for instance, he comments on the spectacular character that the nation acquires in such events:

> There are many young people with red bands around their heads. National flags wave around. The Wave [la Ola] is interminable and precise, a wonder of discipline. On the ring-side (the Golden Zone) those who sold everything to get a ticket (900 dollars) talk to each other with illuminated eyes. It is the UNIQUE EXPERIENCE, and here vehemence eclipses itself. México! México! and only by shouting can one sense the din of the words. The TV catches the public in its delirium and the public delirates for the exclusive benefit of the TV.

(ibid.: 26)

During the same match, he describes the public as it watches itself on screens around the ring, or as they call their friends and family on their mobile phones to describe the experience of being there. The thrill of the multitude, for Monsiváis, is to be able witness an event, to be able to say that one was there.

In the last text in the book, Monsiváis imagines the apocalypse in a futurist Mexico City so big that it occupies almost half of the country (from Guadalajara to Oaxaca). In mock Biblical language, he declares to have seen the Beast: 'And the people applauded it and took photos of it, and recorded its exclusive interviews while, with a clarity that would become painful fuzziness, I became aware of the knowledge of posterity: the worst nightmare is the one that definitively excludes us' (ibid.: 250). Hence, his definition of Mexico City as 'post-apocalyptic':

the worst already happened (and the worst is the monstrous population whose growth cannot be stopped), and yet the city works in a way that for the majority seems inexplicable, and each person extracts from chaos the compensations that somehow balance the sensations that life is unlivable.

(ibid.: 21)

In other words, for Monsiváis the very condition of being a 'megalopolis' (and one of the largest cities in the world), becomes, in the case of Mexico, a spectacle. The national pride expressed in 'there is no country like Mexico' has been converted, according to him, into pride in 'catastrophe, and demographic explosion', that is, the pride in declaring that 'Mexico is the most populous city in the world' and 'Mexico is the most polluted city in the world' (ibid.: 19). As Muñoz observes, 'Monsiváis creates a disproportionate metaphor – the post-Apocalypse – to describe a situation that is equally disproportionate – Mexico, the megacity' (2003: 93). But his metaphor also points to the spectacular aspect of the statistical claims. In the prologue to the book, Monsiváis explains, with Debordian overtones, that thanks to the spectacle 'disorder quietens down, multitudes accept the discipline of awe and a perfect mixture of autocratic imposition and democratic leveling can take place' (1995: 16). The spectacle, then, brings (suppressive) order and control to the chaos of the megalopolis. Attenuating his own statement, he also goes on to add nonetheless that 'genuine fun (irony, humor, irreverence)' escapes the control of the spectacle and consumption and can show that 'in spite of everything, some of the rituals of chaos can be a liberating force' (ibid.: 16).

Watching the multitudes that engage in the 'rituals of chaos' the *cronista/flâneur* maintains his position of distant, if at times sympathetic, analyst, of the one who knows that the spectacle can be, as he says in the prologue, a form of control. Popular Christianity (Guadalupismo), particularly, is seen by him to be the result of 'poverty, an understanding of the world through ritual acts, a lack of support' (ibid.: 40). Television multiplies the millions who participate in the celebrations of Guadalupe into many more millions, provoking him to say that: 'Someone asks himself: will they end up looking at the set as if they were attending Mass? The sceptic answers: Will they end up immersed in Mass as if they were watching TV? . . . On television the multitude belongs to the spectacle in a way never predicted by the temples' (ibid.: 47). The distance that separates him from the multitudes that follow the rituals of Guadalupismo is, as we can see, abysmal. Yet, he manages to attenuate it by turning his *flâneur* eyes on to the commentators on popular religion ('I observe the observers of popular religion' (ibid.: 50)) whose prejudicial comments are not dissimilar to his own. And above all, by ending his series of *crónicas* on popular Christianity with a list of anonymous popular 'quotes' that reveal irreverence and humour from the believers themselves.

Throughout the *crónicas*, Monsiváis carefully punctuates his critical views of the multitude and their rituals with ironic self-critiques of his own limits and his incapacity to join in. In the third *crónica*, for instance, he does so by creating a character, Juan Gustavo Lepe, who goes to the place where national football victories are traditionally celebrated (the 'Ángel', the Independence monument on Paseo de la Reforma, in the centre of the city) a few days after watching a debate about football on television. A fanatic lover of football who remembers details of every game, Juan Gustavo still feels disturbed by the declarations of some participants in the debate, who somewhat coldly had analysed the passions felt by the fans as 'mechanized carnivalization'. As a result, although he is present at the celebration, he cannot actually take part in it, because new forms of analysis keep interfering with his happiness. What we see then is a comical fluctuation between the desire to join the multitude and the contrary, conflicting urge dispassionately to dissect and analyse its (and his own) reactions. Juan Gustavo is thus a projection of the author's own role as *flâneur*.

Later in the book, Monsiváis more directly criticizes himself for not being able to join in the multitude. Watching a concert by the young pop idol Luíz Miguel, he comments: 'Ballads mass-produced for concerts that are mass-produced in order to create mass-produced reactions. I mumble this from my generational and class isolation. And, without raising my voice, I get mad at myself' (ibid.: 191). It is partly thanks to his class (understood here in terms of birth privilege and money but also, and above all, as access to good education), that the narrator cannot, as a *flâneur*, become part of the multitude. Benjamin described the *flâneur* as standing 'on the threshold of the metropolis and of the middle class. Neither has him in its power yet. In neither is he at home. He seeks refuge in the crowd' (1999a: 10). According to his own account, *flâneur* that he is in many respects, Monsiváis knows that his class separates him from the multitude, where he cannot hope to find refuge.

Meanwhile, the sheer size of the megalopolis and its great social contrasts have created other forms of *flânerie*, besides the engagement of Ramírez and the haunted detachment of Monsiváis. In Mexico, on her television programme *Aquí nos tocó vivir* (This is where we happen to live), Cristina Pacheco has been interviewing people from the most marginalized sectors in different parts of Mexico City for no less than 29 years. (In the last few years, the programme has been extended to draw in interviewees from other regions in the country.) Aired twice a week by 'Canal Once', a public television channel, the programme boasts of 'giving voice' to those who traditionally have none (http://www.oncetv.ipn.mx). Intensely Mexican in feel and impossible to translate (at least into English or Portuguese), the name of the programme already sets the tone since in Spanish it gathers all of 'us' (nos) in a first-person plural which grammatically proves to be the object of greater forces well exemplified in the megalopolis itself.

Accompanied by a cameraman (currently David Segovia), Pacheco travels to the houses of her interviewees or their workplaces (most often the house doubles as the workplace) and talks to them and their families about their work and their lives. The act of 'giving voice' is physically enacted by the presence of a bigger-than-necessary black microphone that moves between Pacheco and her interviewees. Learning of injustice and exploitation or worse, she rarely if ever offers outright judgment, a practice that in effect makes her weekly quests the more telling as critical insight into 'la gran familia mexicana', the national 'happy family' cynically vaunted in government and commercial propaganda.

Additionally, Pacheco usually wears black or dark/neutral colors, leaving pronounced colour to characterize the interviewees' homes and clothes. The camera work makes ample use of close-ups, giving particular emphasis to the interviewees' faces. Pacheco often embraces or touches her interviewees, and in any case situates herself very close to them, to make them feel comfortable but also to approach them socially, to put herself physically at their level, as it were.

Commentators on the programme usually insist that it allows viewers to know how poor people in very different parts of the city live. Fernando Benítez, for instance, starts a short review by saying:

> That's right, we happen to live here ('aquí nos tocó vivir'), but we don't know the lives of millions of people in the place where we happen to live. Cristina Pacheco shows them to us. We follow Cristina's slim and sinuous body as she, microphone in her hand, has the talent to talk to street kids, drug addicts, to the blind who toil reading of Braille to support their families. She talks to second-hand booksellers, with those who write letters for lovers at the doors of Santo Domingo church; with those who live in tenements semi-destroyed by the 1985 earthquake; with the street comedians and vendors; with prostitutes . . . ; with some of the five million faithful who on the 12th of December come to the shrine formerly called Villa de Guadalupe.
>
> (n.p.)

The fact that the viewers and interviewees share a common space (Mexico City) is an important point in this (and other) articles about the programme. According to it, Pacheco renders visible the invisible majority of the city,[2] and makes its different inhabitants know (of) each other. Who are, then, the programme's implied viewers? The 'nos' (us) of *Aquí nos tocó vivir* refers to all of those who live in Mexico City (more recently, in Mexico as a whole, though the programme is still mostly centred in the city and is still perceived as a programme about the city), as well as to the poor, whose houses are the ones featured on it. The idea of a shared space, quoted by Benítez, is in fact intrinsic to the programme's initial title.

Yet, as a matter of fact, Canal Once is not a popular channel; most of its viewers have higher-than-average educational levels, though not necessarily university degrees. So, although *Aquí nos tocó vivir* is among the highest rated programmes on the Channel, its most likely viewers are the reasonably educated middle classes. We can probably say therefore that, for the most part, the programme brings the lives of disadvantaged inhabitants of Mexico City into the homes of less disadvantaged middle classes. And although its explicit sentimentality and pathos may not make it a favourite among arty intellectuals, the following declaration by the media critic Fernando Báez Rodríguez makes it clear that it is difficult to predict exactly who the viewers of *Aquí nos tocó vivir* are:

> The [first of the] three programmes with highest rating in Canal Once, in the last few years, has been 'Aquí nos tocó vivir', the multifaceted programme by Cristina Pacheco. You can accuse her of populism – she is always 'so understanding', and will take the side of the poor at any moment – but you can't accuse her of not allowing the people to speak or of making a programme that is not interesting. Those who have never stuck with 'Aquí nos tocó vivir' when zapping through TV channels may throw the first stone.
>
> (n.p.)

It is, of course, technically incorrect to refer to Pacheco as a *flâneur* – one can hardly compare the highly prepared media interview scheme that invades someone's house to the chance encounters of walking in the city. Yet, the act of casually seeing Pacheco's interviewees on television through zapping may, all differences considered, be called a kind of contemporary *flânerie*, where 'citizens of leisure' can be unexpectedly made to observe inhabitants of their own city whom they otherwise would probably never meet.

URBENAUTA

Back in São Paulo, the act of moving through the city and observing its inhabitants has recently taken an extreme turn through the work of Eduardo Emílio Fenianos, a self-proclaimed *urbenauta*, a term the author quaintly derives from *astronauta*. In 2001 he undertook a much publicized 'expedition' through the city of São Paulo. The travelogue that resulted from it appeared as a book in 2002 under the title *São Paulo. Uma Aventura Radical* (São Paulo: A Radical Adventure). Parts of the expedition were also shown on the Globo television programme *Fantástico*, which airs at prime time on Sundays. Fenianos, a journalist, claims to have spent a whole year preparing for the journey, which started in February 2001 and lasted 120 days.

The 270-page book is well produced, though the cover rather simple-mindedly stresses the analogy between jungle and city implicit in the author's characterization of his adventure as an expedition. Superimposed on a dim land map adorned in sixteenth-century woodcut style with naked savages and wild fauna, his jeep keeps company with a toucan, skyscrapers and a naked native child holding a modern football. Inside the covers, there follows a coloured satellite map of the city that delineates its five zones, the centre and the outgrowths in the four directions. After a brief explanation of the project,

further colour images portray the vehicle used by the *urbenauta* and the clothes
he wears. Inspired by astronaut gear, the clothes include silver-lined rubber
suits which, according to the description, would allow the traveller, when
necessary, to swim unscathed in the extremely polluted waters of São Paulo
rivers; and brightly coloured canvas suits that 'would call so much attention
that nobody would dare to harm him' (2002: 8). The vehicle was a brightly
painted SUV named (after Swift's traveller) 'Urbenave Gulliver', equipped with
camera, external speakers to 'communicate with the population' (ibid.: 9),
GPS system, compass, cellular phone, and organic and non-organic rubbish
bins that 'could also be used by the population, as a way of educating them'
(ibid.: 9). For the first phase of the expedition, Fenianos invited a group of
military men to join him from the Pelican Squadron, normally based in Mato
Grosso do Sul and used to travelling through jungle terrain.

The expedition was designed to cover the city in zones and was divided
into two parts or phases, named 'Selva' (Jungle), and 'Selva de Pedra' (Stone
Jungle). The first phase lasted 13 days, during which Fenianos, with the
help of the military, travelled through the non-urban parts of the municipality
of São Paulo: the Cantareira mountain range and park to the North of the
city, and to the south, Guarapiranga (a suitably native name) where the dam
that provides drinking water for the city is located. In phase one, the author
and his military companions were supposed to instil the expeditionary ethic,
toughening up as they navigated the waters of the city's main rivers, including its polluted urban rivers, taking in the celebrated Riacho do Ipiranga, a
historical marker for the nation on whose banks Brazilian Independence was
proclaimed. During phase two, Fenianos was supposed to visit the 96 urban
districts in São Paulo's five zones.

The expedition set itself the following rules:

- whatever happens, not to go back home.
- never to sleep in the *urbenave*.
- stay at least one day in each of the apportioned districts.
- in the first phase, eat and sleep in the forest and on the banks of
  the rivers to be explored.
- in the second phase, find shelter and food in the district of the day,
  preferably in people's homes, in order to establish contact with the
  population and enter their world; and
- of the 120 nights, only 5% could be spent in hostels or hotels and
  these six nights should be for taking stock, drafting reports, and re-
  organizing of the route upon passing from one zone to the next.

(ibid.: 7)

In other words, the expeditionary model was to be enforced with the precision
of a military campaign, especially in the quantification of time and space.

After the map, the description of the gear and the rules, the book moves on
to a 'Prefácio Interessantíssimo' (Very interesting preface). This is a blatant

reference to the preface under the same name that Mário de Andrade had put in the collection of his *modernista* poetry *Paulicéia Desvairada* in 1922, the year of São Paulo's much noted Semana de Arte Moderna. Fenianos is happy to re-use the title but shies away from any association with the two *paulista* modernista poets, Mário and Oswald de Andrade, by saying that his São Paulo 'is not Oswald and Mário's city. It is Christopher Columbus's and Charles Darwin's' (2002: 11). With this disavowal, he wishes to insulate his book from connections with the sedentary world of literature, rooting his expedition rather in the realms of science and exploration. At all events, he disregards what Columbus's voyage actually came to mean for America's first peoples, along with Darwin's often problematic categorizations of them.

The author's anxiety can be explained precisely by the fact that *São Paulo. Uma Aventura Radical* is in fact and in all senses, closer to literary invention than to the scientific and exploratory modes it says it wants to emulate. First, because the expedition's most conspicuous product is not a scientific discovery or theory, model, or a conquering of territories, but a book. Second, because as a book it cannot escape the fact that the comparison between the (modern) city and the jungle has well-known literary antecedents, not least among urban naturalists like Zola or Upton Sinclair. As Roger Caillois remarks, in a passage famously quoted by Walter Benjamin with reference to the detective novel:

> We must take as an established fact that this metamorphosis of the city [in the detective novel] is due to a transposition of the setting – namely from the savannah and forest of Fenimore Cooper, where every broken branch signifies a worry or a hope, where every tree trunk hides an enemy rifle or the bow of an invisible or silent avenger. Beginning with Balzac, all writers have clearly recorded this debt and faithfully rendered to Cooper what they owed him.
>
> (Benjamin 1999a: 439)

It is the widespread knowledge of literary and cinematic models of adventure and exploration (which if we accept Caillois's and Benjamin's claims derive from Cooper) that make Fenianos's expedition financially viable and endow its imaginary and rhetoric. Fenianos, the military contingent, the media channels that were willing to sponsor him, the people who had heard of him on the radio or seen him on television and wished to make contact and to offer him their support and shelter – all of them *knew* that São Paulo was not a jungle to be explored by someone driving a safari-type SUV and wearing astronaut clothes. If they accepted to be involved in it, it was because the implicit and explicit assumptions of the expedition (and the analogy insistently drawn between the jungle and the city) somehow made sense to them, or, in the case of the media, offered the chance of generating audiences, and money.

What are these implicit and explicit comparisons? First, and most obviously, that the city is *unknown*. This is the basic assumption in Cooper's literary model, in the detective novel that transposes it on to the city, and in the stories of adventure and exploration that feed into Fenianos's expedition. On the website of the expedition, the journalist describes the logic behind the trip precisely in terms of the unknown: 'Knowing it [São Paulo] deeply can happen only through a journey or expedition organized along the same lines as the expeditions led by Marco Polo, Christopher Columbus or the Villas Boas brothers who explored the interior of Brazil from the 1940s on' (n.p).

Curiously, São Paulo was not the first exploration undertaken by the *urbenauta*. Before attempting his great expedition through the megalopolis, Fenianos (a *paulistano*) had made a pilot trip of Curitiba, a much smaller city in the South, and his website also mentions a subsequent excursion to Florianópolis. São Paulo, however, was the only adventure to be truly successful in the series, and the only one to produce a full-length book, to receive plenty of sponsorship and media coverage, and so forth. The reason for this is that of all the cities involved, São Paulo is the only one that plausibly can come across as being unknowable, as being so big as to be rendered absurdly beyond cognition, and therefore the only one that could be subjected to the absurdity, as it were, of a SUV exploration complete with military consultants, special equipment and the rest. In other words, although Fenianos's expedition (and the book that results from it) does make certain aspects of the city better known, it does not create a new corpus of systematic knowledge about São Paulo. What it does, however, is to highlight the fact that the megalopolis is, by definition, unknowable. The colourful and safari-like vehicle, the astronaut clothes, the outside speakers that tell cars to move out of the way – they are all performative demonstrations that the city is like an unknown jungle. They enact for the inhabitants of São Paulo what they are already aware of: that the megalopolis is too big for anyone to know it in the span of a normal life.

The second element of comparison between the megalopolis and the jungle that is made evident by the expedition is *danger*. Once again, the SUV and the special clothes are not exactly the best solution for the problems of urban danger. But by highlighting the comparison between the dangers of the megalopolis and the dangers faced by the members of a jungle expedition, they perform an image of the city (the dangerous jungle) that appeals to its inhabitants. They find it amusing because they are already all too familiar with the idea, though it can materialize into terminal damage. As Fenianos reminds us, immersion in the polluted rivers can be expected to result in death, a concept integral to the plot of the film *16060* (1995), Vinicius Mainardi's whimsical disquisition on relationships between the very rich and the very poor in São Paulo.

In the context of the daring adventure into the perilous unknown, other elements having to do with the expedition itself begin to signify: how the act of moving through the city can be made into adventure, and how the traveller can become, as he engages in the adventure, a hero of sorts. All of those aspects of Fenianos's expedition were already part of the of the *flâneur*'s *raison d'être*: moving through an unknown and possibly dangerous city, finding adventures and becoming a hero explorer, epic in his way. Baudelaire talks about urban heroes in his short text 'Heroism of modern life', which was included in *Art in Paris* alongside one of his *Salon* reviews:

> The spectacle of fashionable life and the thousands of floating existences – criminal and kept women – that drift about in the underworlds of a great city: the *Gazette des Tribunaux* and the *Moniteur* all prove to us that we need only open our eyes to recognize our heroism.
>
> (1965: 32)[3]

In Baudelaire's opinion, the Parisian heroes of his day did not need epic paraphernalia: grey suits worked well enough, as 'the necessary garb of our suffering age, which wears the symbol of our perpetual mourning on its thin black shoulders' (ibid.: 118). In a São Paulo that has grown out of all proportion and has become exponentially much more dangerous, heroic *flânerie* has to become, like so much else, a grandiose media spectacle.

Fenianos is not a great writer, and *São Paulo: uma Aventura Radical* is riddled with repetition and commonplaces. What makes the book worth reading is the freshness with which the author tries to leave his class privileges behind when embarking on this super-protected, media-fuelled adventure. One of the rules listed at the beginning of the book is that no matter what happens, he could not go home, and he explains that 'home' for him is the upper-middle and upper-class neighbourhood of Alto de Pinheiros. When he arrives in one of the poor neighbourhoods in Zona Sul (South side), someone he meets tells him that true heroism is to be able to live on the Brazilian monthly minimum wage. From then on, Fenianos decides to incorporate this further challenge into the concept of his expedition: he tries to live for a month on the minimum wage. Needless to say, the experiment is contrived: he is being put up in people's houses anyway and already has a great amount of gear in his possession, so that he is not really 'living', in any normal way, on that amount of money. Throughout the length of the expedition, the lessons that he learns and tells his readers have mostly to do with the fact that one can only get to know the city and its people if one leaves behind all class privileges and approaches people – especially the poor – in a humble way. SUVs and astronaut clothes would hardly have been necessary to bring about this message – except perhaps if one lives in a city like São Paulo, where the class abyss has acquired colossal proportions and elicits spectacular response.

Still, at certain moments in the travelogue he shows great sensitivity, lamenting that parts of his own city should appear to him as completely strange and unknown. Such is his reaction when he arrives in Jardim Ângela, one of *bairros* in Zona Sul, in the region of Capão Redondo. Listening to local rap on the music player of his *urbenave*, he paints for us a picture of the *bairro*:

São Paulo is beginning to make it clear to me how it is that it manages to have millions of inhabitants. There is a predominance of *favelas*, unfinished houses with few rooms where parents, grandparents, children, nephews, and uncles all cram together as if they were in a bird's nest. Vehicles pass which make me wonder how they move at all. They are old Variants with no doors, decrepit Brasílias (a Volkswagen model), and even horse-drawn carts. Horses graze beside cows while the tyre man changes two more tyres and two lads wait for the van that will take them to the centre of the city where they hope to find a girl. When I ask people what is there to do here, the majority of them say 'nothing'.

There are lots of people on the streets. An amount I have never seen in the city's more central or wealthier streets. In just a few days in the *periferia* I can already conclude that the street here is not just a means of locomotion: it is a meeting space.

(2002: 56)

He then catches the smell of a barbecue and sees a large group of people in front of a construction site. They tease him: 'Hey, cutie, are you going to war in that big jeep?' And, 'Who's a bit special around here, we may be stoned but look at that outfit.' He asks what is going on and is told it is a *mutirão*, a Tupi-Guarani term which denotes people working as a group to help one of their number complete an arduous task, like building a house, the custom like the word being indigenous in origin. 'We call the neighbours, one tells the other, friends come and friends of the friends of the friends and together we are mixing the mortar to put a second storey on to the house, and we use the occasion to have a barbecue and drink beer' (ibid.: 57). As he observes: 'No engineers, architects, building approvals, inspection reports. In the eyes of the law, just another illegal house in an illegal, invaded piece of land' (ibid.: 57). As a young man from the wealthy side of the city, he is witnessing for the first time the situation described in the analyses by Caldeira and Maricato: the gap between the regulated wealthy city and the unregulated, illegal occupation of the *periferia*. Fenianos offers to help serve the barbecue and is then able to talk to the people, and find a place to stay. Before setting off on his journey he had been warned against going to places like Jardim Ângela and Capão Redondo because they were extremely dangerous, which prompts him to comment, ironically: 'Nobody had told me or warned me about this Jardim Ângela: "Hey, be careful, you'll find a *mutirão* in Jardim Ângela and eat a great barbecue and make friends"' (ibid.: 58).

As the night wears on, the locals become more expressive about their own feelings towards the wealthier parts of the city where he comes from:

From here to Avenida Paulista [the financial centre] it is not even 20 Km, but why don't you ask a businessman if he has the guts to get into his car and come here. Rich people only come to the *favela* to buy drugs. I bet everyone says you're insane to be in a place like this, where there are only thieves and blacks. They see the *periferia* on television and don't even know what it is. Why don't they come and see it with their own eyes?

(ibid.: 59)

He is then given a tour of Capão Redondo in his *urbenave*, and is shown that 'not all the streets are empty they make them seem on television news in order to prove that everyone is scared' (ibid.: 61). As he leaves the *bairro* three days later, he receives a call from a friend asking him where he was. 'I answered I was leaving Campo Limpo, Capão Redondo, and Jardim Ângela. "Didn't they eat you? Are you still alive?" The same prejudice I had before going there' (ibid.: 63).

This touching admission does much to lend credibility and worth to the urbenauta's account of his exploit, self-promoting, faintly comic if not plainly ridiculous as it might otherwise seem. Self-defined as a *flâneur*, though motorized, Fenianos succeeds in revealing a side of life in fabled Capão Redondo that is not always apparent even in inside accounts (see p. 107).

The experience of *flânerie* in contemporary Mexico City and São Paulo is strongly marked, as we could see in all of the examples above, by a strong awareness of class difference. There is of course nothing new about this fact in itself. Many of Baudelaire's poems address the issue of poverty – 'The Eyes of the Poor' being probably the most notable example, because it shows two opposite behaviours: the poet's sympathy and compassion, and his female companion's contempt. As Marshall Berman observed, one has to be struck by 'Baudelaire's sympathy and generosity, so different from the standard image of an avant-garde snob who exudes nothing but scorn for ordinary people and their travails' (1988: 144). In the case of the the two Latin American megalopolis, sympathy and generosity are clearly not enough to deal with the immense social gap. In Armando Ramírez's stories, police brutality turns the mere act of walking in the city into an evidence of class privilege, so much so that the *cronista* who observes the poor walker has to show constant awareness of his limits, i.e. the limits imposed on *flânerie* by his class and superior education. The same happens to Carlos Monsiváis when he observes the crowd: he is deeply and permanently aware that he cannot join in, for reasons of class and education. Cristina Pacheco's television program is perhaps the one that gets the closest to unguarded expressions of sympathy and generosity, though the very fact that it has been entertaining audiences once a week for almost thirty years, with its glimpses of the megalopolitan 'other', says much about the size and pervasiveness of class difference in that city. In Martinelli's film *Opressão*, it is not only poverty, but gender, that makes walking difficult and fatal. And in Fenianos's *São Paulo: uma Aventura Radical*, sympathy and generosity are mediated by the absurdity of the paraphernalia that, whether he likes it or not, freezes the upper-class *flâneur* into a role of colonizer, or alien.

II.

Find your place in the neighbourhood

4 Barrio/bairro

Within any given city, only a few neighbourhoods become famous, or infamous, to outsiders. These are mostly extremely rich or very poor, physically striking or have something remarkable about them, culturally speaking. No-one has to be from New York, for instance, to have heard of Harlem, Bronx, Chinatown, or the SoHo – neighbourhoods variously marked by histories of violence and racial conflict, ethnic autonomy, art or trendy bohemia. Most often, these are also the neighbourhoods that are seen to project the strongest cultural profile – either because something remarkable about them has attracted the attention of writers and artists, or the converse, that is, because cultural representation (not to mention tourist or real estate propaganda) has made them better known to begin with.

In Latin America, the closest equivalent to neighbourhood is the term *barrio* or *bairro*, which has a (possibly Arabic) common root though not quite the same resonance in Spanish and Portuguese. In the cases of Mexico and São Paulo, most cultural representations of their *barrios* and *bairros* have tended to favour and concentrate on those that are poor or impoverished. In scholarship, these have been the parts of the Latin American megalopolis studied by anthropologists and sociologists, no doubt because they are the places perceived as having a certain cultural autonomy, some form of *identidad barrial* (neighbourhood identity), or as epitomizing key qualities of city life there. This is certainly the case of the two neighbourhoods that will be considered here as places to 'live in the city': Tepito, a central *barrio* of Mexico City, long renowned for its resilience; and Capão Redondo, a *bairro* on the outskirts of São Paulo that through music and culture has in recent years come to impinge strongly on the image of that city.

The English word used here to translate *barrio/bairro*, neighbourhood, has been authoritatively traced back (in the *Oxford English Dictionary*) to medieval understandings of friendly relations and feelings between people who live near or next to each other, and hence 'the people living near to a certain place or within a certain range, neighbours; a community, a certain number of people who live close together'. Only later (late seventeenth century) is the term spatialized primarily to denote 'A district or portion of a town, city or country, esp. considered in reference to its inhabitants; a small but relatively self-contained sector of a larger urban area'. Though the words *barrio* and *bairro* bear definitions similar to these versions of 'neighbourhood', their connotations vary slightly. In Mexico, the meaning is closer to the first OED meaning, that is, it refers most often to a 'community', and quite frequently to more traditional communities, ethnically marked as having indigenous heritage. The word *colonia* tends be used, in fact, more frequently than *barrio*, to denote a more or less neutral division within the city. In São Paulo, *bairro* bears both connotations, but is perhaps closer to the later OED definition. It may be applied to fashionable and desirable parts of the city no less than to poorer areas.

The maps and charts of officialdom may inform us about the boundaries of a given neighbourhood, and these can be linked in turn to postal codes, school catchment areas, building zones, property value, traffic density, availability of services, and many other factors, yet there is probably no neighbourhood anywhere, and certainly not in Mexico City or São Paulo, whose boundaries are perceived exactly in the same way by all the people who live in it, let alone by outsiders. And neighbourhoods, of course, change a lot with time, perhaps the quicker according to overall rates of growth. So when we talk about neighbourhoods here, we are talking about areas of the city whose definition is reasonably fluid, ever-changing. Yet they are parts of the city that are recognized by a certain name, and around which a certain imaginary has been created. In the case of the megalopolis this is saliently so with Tepito and Capão Redondo. Nowhere is this notion of barrio as such more succinctly exemplified than in the emergence in Mexico City 15 years or so ago of the character and phenomenon known as Superbarrio, a 'Latin American version of Superman' who, garishly attired in his cloak, sides with the poor in their street demonstrations and struggle for justice.

Anthropologists and sociologists, especially in Mexico, have dedicated a lot of time to the subject of *identidad barrial*, seeing it preferentially in the context of social movements. What interests these scholars is the kinds of networks that people living near each other may develop as strategies for survival and political action. As the historian Aréchiga Córdoba points out, having this approach in common neighbourhood studies are nonetheless motivated by quite different interests, and these interests tend to coincide with those behind most artistic representation of neighbourhood.

First, many studies have encouraged (and been encouraged by) an idea of neighbourhood which is inseparable from nostalgia. (This concept plays into the argument in accounts of the city prior to the emergence of cultural studies as such in Britain, for example, in Raymond Williams's work, or in Richard Hoggart's *The Uses of Literacy* (1957), with its diagnosis of life, feeling and language in the wastelands of the once industrial North of England.) In this line, neighbourhoods have over time come to be treated as enclaves of tradition caught within a generally modernized (and degenerate) city. In such studies, neighbourhoods are perceived either as images of what the city used to be like, or as a kind of intermediary between the city and the country:

a place that resists modernization because of inherited and remembered custom (Nivón 1993: 16). In cinema, a recent example of this version of neighbourhood is Jorge Fons's *Callejón de los Milagros* (1995) where 'miracles' are limited to a cul-de-sac (*callejón*) in the traditional heart of the city, discussed at length below with respect to Tepito. Based on a script by Vicente Leñero (based in turn on the novel of the same title by the Egyptian Nobel Prize-winner, Mahfouz Naguib, that correlates Cairo and pyramids), the film by Fons adapts well to Mexican terrain the conflict between modernization and *barrio* tradition, exacerbated in the story of the local girl who is lured into prostitution. Tradition, in this sense, may also have played into attempts to read Tepito, but never Capão Redondo.

Difference of the neighbourhood from the rest of the city has also been repeatedly ascribed not so much to time depth as to ethnic variation, to the preponderance in it of people who look and sound different, who to this degree may seem exotic, possessed of cultural codes that are incomprehensible to a majority of others. *Barrio* was first used in English in just this way, to locate the 'Spanish-speaking quarter' of a US city. It similarly marks the birth of *bairro* literature in São Paulo, in works by Juó Bananére and Antônio de Alcântara Machado which focus on districts noted for their Italian immigrants. In Mexico City, *barrios* distinguishable on such grounds appear to have played less of a cultural role, perhaps for reasons inseparable from the country's longer and denser demographic history. In any case, this order of difference cannot fail to be problematic from the start, effectively depending as it does on the highly sensitive concepts of race, language and culture.

Conversely, other scholars have wanted to see the neighbourhood not as different from the rest of the city but as the epitome of it, a microcosm that concentrates its characteristics and which thanks to its contained limits is easier to study or represent (an academic consideration of some practical weight). In other words: the neighbourhood in these cases comes to stand for the whole city, and sometimes – notably in Mexico – the whole country. Some classic accounts of neighbourhoods fall into this category, and probably most studies of cities share this characteristic, at least to a certain degree. Oscar Lewis's world-famous monograph on Tepito, *The Children of Sanchez* (1961), is an excellent example, and much apropos here. Though specifically about a large *vecindad* (collective housing area) in Tepito, *The Children of Sanchez* represents what Lewis called 'culture of poverty'. Having had a major impact in its day, this memorable term can be (and indeed often has been) applied to other parts of Mexico City, of the country, and of the world as a whole. The same might be said about versions of the *barrio* subscribed to in films made in the same post-war period. A leading example is Ismael Rodríguez's melodrama *Nosotros los pobres* (1947), where the poor *barrio* stands for the whole city, in its power to corrupt and degrade country folk who not long ago have arrived from the provinces.

A more recent literary example of this angle on neighbourhood is Juan Villoro's *El disparo de Argón* (1991), set entirely in the neighbourhood of San

Lorenzo, in the north-east of Mexico City. In spite of the fact that urban pollution, traffic, and general decay are ever-present in the novel, the plot itself could well have been set in a small town. The characters all know each other and recognize each other on the streets. There are local workers and professionals, like the barber, the doctor, and so on, who greatly resemble those found in narratives set in smaller places. Yet again, the *barrio* San Lorenzo stands for the city, and the city, in the case of Mexico, as we have suggested and as this novel constantly insinuates, largely represents political power at the national level. In other words, San Lorenzo is a metonymy of Mexico.

Evidently neither sociological studies nor artistic representations of cities will necessarily limit themselves to one or other of these models, concentrating on how things were or ethnic difference, or seeing the neighbourhood as microcosm. Rather, we are likely to see combinations of more than one, or oscillations between them. In any case, both the social science model of neighbourhoods and cultural representations of them tend to focus on the relationships between the people living in them, the human networks inherent in them.

Contemplating the city in just these terms, Manuel Castells (1989) identifies it as a 'dual city', pointing to the acute social division between two types of communities. On the one hand, there are the elites, who thanks to their social advantage are able to establish network connections with the whole world, being a world class in this special sense (capable of arming and protecting itself at home where need be against those less fortunate). On the other, there are the poor and displaced, harassed and hemmed in, who develop internal networks in order to survive at all (ibid.: 227–8).

Acknowledging the same or similar evidence, Nestor García Canclini (an Argentinean sociologist long resident in Mexico City) goes in the other direction. He argues rather that the poorer and traditional neighbourhood is being transformed out of recognition, since its human networks are being updated and replaced with new types of relationships. According to him, this is because: 'social identification is more and more based on semiotic models provided by the culture industry rather than on the signifying structures or the temporality of the neighbourhood' (1998: 27). For her part, Rossana Reguillo, in her accounts of youth culture further attenuates the concept of neighbourhood when she says that for the youth of late twentieth century, 'the *barrio* understood as their own territory is no longer the centre of the planet' (2000: 142–3).

Recent cultural production from Tepito and Capão Redondo allow us to think of neighbourhood networks in yet another way. In his recent work, the novelist and *cronista* of Tepito, Armando Ramírez emphasizes an idea of neighbourhood that is intensely local yet fully charged at the same time through its connection with global modes of production and consumption. In itself, this obviously impinges on the rich–poor dichotomy respected by Castells and refined by Canclini, whatever their other difference may be. In the same vein, rappers and writers from Capão Redondo are forging forms of social identity that are intensely local while belonging, at the same time and thanks to current electronics and global economic processes, to international networks that are sensitive to contemporary youth culture.

What is perhaps most important, and new, about these two phenomena – the word from Tepito and Capão Redondo, as it were – is that, in contrast to most portrayals of neighbourhood hitherto, they come and are produced from within. Studies of *barrios* and *bairros*, as well as literature and films about them, had been mainly the work of outsiders.

In the late nineteenth century, fascinated by the massive urban change of his day, Émile Zola produced accounts of the city that have been foundational for literary narrative, not least his own Naturalist novels. His 'ethnographic' method is well known: reportedly he would go to the neighbourhoods and places he wanted to write about, notebook in hand, and interview people. He would make plenty of notes, on the basis of which, once back home, he would attempt to reconstruct local life in his own narratives, animating appropriate characters.

By contrast, Armando Ramírez and the rappers and writers to be discussed here are all locals, and strongly committed to the place they live in, within the megalopolis. There are, it is true, differences between these witnesses from Tepito and Capão Redondo. Ramírez went to university and has been a professional writer for many years (he writes newspaper columns, blogs, film scripts as well as novels), while the rappers and artists from Capão Redondo have had very little formal education. Like the Capão Redondo artists, however, Ramírez comes from humble origins, and his work has been always committed to the *barrio*: its language, the everyday experiences of its citizens, its characters, his own experiences living there. When asked why Tepito identity had become so important in his literature, he answered because he came from the very heart and centre of Tepito. Hence, from childhood on, he was too well known to lie at school, or pretend like many of his friends that he lived somewhere else, in the *barrio* next door, for example (Rojas 2006: n.p.). Like the young men from Capão Redondo, then, Ramírez's strong sense of *barrio* identity is, at least in part, a reaction to the standard pre-judices against the poor who live there, a defiant response. His books, especially the first ones, have often been referred to as '*barrio crónicas*', that is, rather down-market neighbourhood stories, and for many years they were not considered to be 'serious' literature. Even now he continues to be a relative outsider in Mexico City's literary circles, the object of a curious distancing, and under-recognized.

The *barrio/bairro* pertinent to the works to be considered here, moreover, appeals to an idea of space production, itself worth noting. All the works studied here appeal to an idea of *space production* that closely resembles Henri Lefebvre's in his seminal *The Production of Space* (1991). They share with the French philosopher a non-Euclidean notion of space, seeing it not as an absolute or neutral 'container' where humans or objects locate them-selves, but as an integral part of human activity. Defending the need for his space-centred approach, Lefebvre takes as a point of departure Marx's replacement of the 'study of things taken "in themselves", in isolation from one another, with a critical analysis of productive activity itself (social labour; the relations and mode of production)' (ibid.: 89). He then goes on to claim that:

> A comparable approach is called for today, an approach which would analyze not things in space but space itself, with a view of uncovering the social relationships embedded in it. The dominant tendency fragments space and cuts it up into pieces . . . Thus, instead of uncovering the social relationships (including class relationships) that are latent in spaces, instead of concentrating our attention on the production of space and the social relationships inherent to it – relationships which introduce specific contradictions into production, so echoing the contradiction between the private ownership of the means of production and the social character of the productive forces – we fall into the trap of treating space as space 'in itself', as space as such.
>
> (ibid.: 89–90)

Neither Tepito in the writings of Armando Ramírez nor Capão Redondo in the songs and texts by local rappers and writers are ever treated as 'space in themselves'. On the contrary, they are described, on the one hand, as the result of political and economic conditions, and, on the other, as 'works in progress', that is, as spaces that are being *produced* by the activities of their inhabitants.

Further, the two neighbourhoods in question, Tepito and Capão Redondo differ quite one from the other in themselves. Each represents, *barrio/bairro* in the megalopolis, a readily recognizable yet diametrically distinct model of urban development, the one traditional and central, the other new and peripheral. Attempts to understand them theoretically differ correspondingly. Beside Tepito with the wealth of studies and champions it has inspired, Capão Redondo is very much the poor relation. It traces no memory to Pre-Columbian days, and as yet has no Oscar Lewis. Nor have the nascent accounts of Capão Redondo's music and culture led so far to its being situated in the longer story of the *bairro* in Brazil. As second-generation immigrants to São Paulo, from the north-east of the country, its inhabitants await comparison with such immigrant predecessors as the Italians in Bixiga, with whom paulista *bairro* history is normally assumed to have begun.

Geographically, Tepito is located in a very central position, north of the Zócalo in the very heart of Mexico, innermost in the inner city. To that degree, it exemplifies the 'inner city' culture of poverty examined by Lewis. At the same time it differs from North American inner cities in the sense that, as we shall see, it claims for itself a very long history, a history that indeed precedes European colonization. It also differs from other inner city models in the sense that it daily attracts thousands of consumers to its markets and street stalls, nowadays better known for the commerce of *fayuca*, illegal imports or pirated goods.

Capão Redondo, on the other hand, is a typical São Paulo *periferia* neighbourhood, that is, somewhere located very far from the centre of São Paulo, in the southern part of the city. Its existence as a neighbourhood dates only to the twentieth century, and it has become well-known in the imaginary of São Paulo in only the past 10–15 years. Its inhabitants are mostly descendants of migrant workers who left the very poor, drought-ridden areas of north-eastern *sertão* to work in the industrialized São Paulo in the 1960s and 1970s.

Nonetheless, both Tepito and Capão Redondo coincide in occupying very similar places in the imaginary of the metropolis they respectively belong to. In the media, they are both billed as dangerous and inhospitable, riddled with thugs, guns and drugs. All the more reason, therefore, to consider what they have to say for and about themselves.

TEPITO AND WHO WAS HERE FIRST

In his collection of *crónicas* about Tepito, Armando Ramírez attributes to the writer Salvador Novo an explanation for this toponym that traces it back to the Nahuatl, the language of the Aztec empire, as a derivative of the term *tepitoyotl*, which means smallness or being small (1983: 14). In Alonso Molina's sixteenth-century *Vocabulario en lengua castellana y mexicana*, tepito 'cosa pequeña o poca cosa', in fact, immediately precedes *tepitoyotl* 'pequeñez'. In other words, it is not obviously absurd for a local history of Tepito to claim its name originated in pre-Cortesian times, reminding everyone that the territory which today corresponds to the barrio was once part of Tlatelolco. As for its size, from the start it appears to have been squeezed between the central pyramids of Tlatelolco and Tenochtitlan just as today it lies within and between Santiago Tlatelolco, Delegación Cuauhtemoc, and the Centro Histórico.

A close neighbour of the Aztec capital Tenochtitlan at the densely populated northern end of the island they shared, Tlatelolco was effectively incorporated into it after its defeat in the late 1400s under the Emperor Axayacatl. Written in the native script (*tlacuilolli*), Aztec tribute lists give a clear idea of the quantity and dazzling variety of goods that regularly poured in from territories often many hundreds of kilometres away, ever more so after Tenochitlan, having dominated the island and the Highland Basin went on to take Coixtlahuaca and the highway to the east in 1467. Though incorporated, Tlatelolco identified itself less than whole-heartedly with Tenochtitlan and in 1520 was said to be where the people turned on Emperor Moctezuma when he was Cortés's hostage and regarded as a traitor. Above all, it very much retained its own market, an institution known in Nahuatl as *tianquiz* that was fundamental in the life of the city and the empire as a whole. The abundant wealth of the Tlatelolco market did not escape the notice of Hernán Cortés when he entered the city. Indeed, the first thing he has to say about it in the report he sent back to Emperor Charles V in 1520 (Segunda

Carta) is the constant buying and selling that goes on in its market plazas, the largest being twice the size of the largest in Salamanca and able to hold more than 60,000 people. He lists the huge variety of products on offer there, brought in from afar in all directions: 'jewels of gold, silver, lead, tin, copper, zinc, stones, bone, shell, sea-shell, feathers', meat, medicines, ceramics, blankets, vegetables, fruits, *pulque* (the alcoholic drink made from maguey), paints, cottons, earthenware, lake and sea fish, game, tortillas, and so on and so on, in obsessive detail, comparing the abundance favourably with anything to be found in Europe, noting the arrangement of goods by 'streets' (*calles*), the judges who settled disputes, and the strict policing of measures of volume and length (Mier y Terán Rocha 2005: 97–8). The Tlatelolco *tianquiz* was immortalized in 1521 as the place where, besieged by Cortés, Cuauhtemoc made his last stand in defending the city island from European attack. Tlatelolco's pyramid is very much still there and after the Revolution became in more than one sense the foundation of the astounding Plaza de las Tres Culturas (the three cultures being pre-Cortesian, early colonial and post-revolutionary).

The devastation wrought on Tenochtitlan-Tlatelolco by the invasion and siege is hard to exaggerate. From the heaps of rubble produced by the invaders' cannon, the area was rebuilt as Ciudad de Mexico, in which Tepito, like other neighbourhoods around it, was openly acknowledged to be a *barrio de indios*. During the first centuries of the colony, the new city as such was reserved for the Spaniards while the *barrios*, or *parcialidades*, as they were officially called by the Spaniards, were the places where Indians lived. With time a new plan or *traza* of the city was drawn up, in which the streets and lineaments of the old Aztec capital were apparent. Freshwater sources were a particular concern, as they had most dramatically been during the siege according to the Tlatelolco Annals (León-Portilla 1992), and Cuauhtemoc's Ordenanza (Valle 2000), being defined now as *acequias*, the Arabic term for the channels that supplanted Aztec aqueducts and ditches. The Spanish colonizers lived inside the limits established by the *acequias*, while the original residents or 'indios' were forced to live outside, instituting what O'Gorman called the 'principle of separation' (1975: 16). The areas where the Indians lived constituted collective property, which in principle could not become part of the system of distribution granted to Spanish colonizers.

The evangelization of the local inhabitants, who in principle lived off limits in the main city, was entrusted to the religious orders (Mier y Terán Rocha 2005: 112) and began with the arrival of the Twelve Franciscans from Rome in 1524. Of course, the limits could never be strictly enforced: the Spaniards soon started to occupy land outside the *acequias*, partly because they feared revolt (Aréchiga Córdoba 2003), while, at the same time, city Spaniards depended on Indian labour and Indian commerce, which implied the daily presence of thousands of Indians within the city limits. In vain, missionaries endeavoured to keep their wards literally on the strait and narrow path to and from colleges and schools of indoctrination in the city.

From the start, the grid plan the Spaniards endeavoured to impose contrasted with the serpentine, untamed occupation of the areas that were then reserved for the Indians.

Towards the end of the colonial period, moves were made in the Bourbon spirit to incorporate the *barrios de indios* into the city proper. After Independence, Indian property in those neighbourhoods began to be systematically expropriated, and a law of 1856 simply made it illegal for collective land tenure to exist in Mexico City and some of its vicinity. Natives could still keep their land as long as they registered it as private property, which many started to do, and there was still recourse to collective legal action. Such possibilities ended by 1868: after executing Maximilian and establishing the Republic, Juarez, himself Indian in origin, proclaimed that all collective land had to go into private hands or become municipal property (Aréchiga Córdoba 2003). This was formally the end of the *barrio de indios*, though, as Andrés Lira observes, not necessarily the end of a way of life or of collective action.[1]

For some historians, like Aréchiga Córdoba, the 1868 law meant for Mexico City Indians, 'the end of the last elements that allowed them to exist in an order that was counterposed to the city. They entered the city as a functional part of its space' (ibid.: 72). It is the beginning, according to the same historian, of a distinct history: the 'naturals' or 'sons of *barrios*' became simply neighbours, devoid of a legal definition that would distinguish them from the rest of the city' (ibid.: 72). Tepitoans tend to disagree, as we will see. Much of the neighbourhood's cultural initiative has sprung from a concept of *barrio* that implies the continuity of native history, a continuity that is denied in such received accounts of the facts. Aréchiga Córdoba himself is ironic about this view of history:

> From then on [1960s], a narrative began to be constructed, with no regard for academic scruple. It traced a line of continuity from the last *tenochcas* and *tlatelolcas*, who fell defending their city from the Spanish invaders, to present-day *tepiteños*. Other arguments in this narrative recover acts of resistance by those who inhabited the *barrio* during the colonial period or independence. What seemed most surprising to me is how this version of history is disseminated among the *barrio* inhabitants.
>
> (ibid.: 35)

Héctor Rosales Ayala shares such a view when he claims that 'it is not possible to find generational continuity from before the end of the nineteenth century. Yet trying to take it back as far as pre-hispanic times remains tempting, since after all the issue is about rights' (1991: 32).

This question continues to be very much at the centre of the discussion of space in Tepito itself, and it will be crucial for our understanding of Armando Ramírez's novels and chronicles. Scholars like Aréchiga-Córdoba and Rosales Ayala seem to understand continuity to mean an unbroken line of descent that can be proven by official documents. Writers and activists like Armando Ramírez take the idea of continuity in broader terms. The direct line of descent or ascent is of course not denied in their discourse. Indeed, if one accepts the general history of Tepito as having been a *barrio indio* until at least the middle of the nineteenth century, how can we imagine that all the Indians who lived there would all of a sudden have disappeared, to be entirely replaced by others of quite another origin? In this vein, they extend the base of their claims citing a general history of occupation of that particular place by Nahuatl-speaking populations from at least the fourteenth century.

Since many, if not most, of the present inhabitants of Tepito are bound to be descendants of the Nahua, their claim need not seem so absurd, at least to me. This is not to deny that the inhabitants of Tepito may be using the notion of Indian heritage to their advantage in order to strengthen their rights to remain in the neighbourhood, or that in that process some of them may make academically suspect or supposedly 'exaggerated' claims. At all events, claims to rights that are based on supposedly exaggerated narratives of heritage, heroism and continuity do not differ greatly from, say, standard European claims to national or regional identity, or for that matter from nationalist narratives generally, not least those of Latin American nation-states.

The modern history of Tepito, from late nineteenth century onwards, centres on a motif of the marginality and criminality which have resulted from new spatial configurations. These have been described by Rosales Ayala:

> the last decades of the XIX century mark the unleashing of new economic forces that would from now on guide the growth of Mexico City, according to the interests of capital and of the governing class of each period. Class differences become more acute and express themselves particularly in the urban sphere: some areas became 'decent', 'hygienic', 'secure', and 'organized' while others were converted into dangerous, indecent, insecure and chaotic territories. Tepito is reborn now under the sign of speculation and the elimination of community property.
>
> (1991: 43)

It is precisely under that guise that Tepito acquired notoriety as a *barrio*. After the Revolution and throughout the twentieth century, plenty of chroniclers, writers and scholars visited Tepito in order to show to people from other areas of the city and from other parts of the country what life was really like in these poor oppressed areas of the capital. *La malhora* (1923), an experimental novel by Mariano Azuela, describes a young woman who, after being brutally raped, becomes a drunkard. In naturalist fashion, the neighbourhood appears in the novel as the medium that condemns human beings to perdition.

A particularly good example of the kind of discourse that the press has constructed about Tepito since the beginning of the twentieth century is the long fictionalized report by Alfonso Lapena and Fernando Reyna. In 1944, Lapena went to the 'neighbourhood of lost souls':

They live under the same cycle, on the same soil as we do. They are men and women that have been or could be like us, but whom crime, addiction, or poverty suffered from the day they were born have made them descend to the condition of infra-human beings, whom society ignores or rejects. There on a street crossing in the very heart of Mexico City, the outlaws, the degenerate, the poor, the persecuted have made their own city, the damned city, the city of hatred, desperation, and hunger.

(quoted in Aréchiga Córdoba 2003: 250)

We can compare it to two examples in the recent press which make similar points: an article in the Spanish daily *El país* and the blogspot '13 radical riders'. After describing all the illegal activities that go on in Tepito, Lonewolf, a collaborator to the latter, advises:

Here is where a multitude of things can happen to you, as soon as you step on to the pavement. A thousand crimes are committed on its streets on a daily basis. Be best advised to stay on the safer side in the Colonias adjoining its territory. Because here in the Barrio Bravo of Tepito – **ASSAULT IS GUARANTEED**.

(Lonewolf's boldface)

The reporter from *El país*, meanwhile, claims that her taxi driver asked to leave her outside the neighbourhood, for fear of entering 'the *barrio bravo de Tepito*', that is the wild neighbourhood (Rico 2006).

The idea of a Tepito not so much lawless as a 'barrio bravo', intrepid to the point of being pugnacious, inspired an early essay by Carlos Monsiváis (1970), which with customary wit, word-play and paradox celebrates the legends that attach to the barrios, not least those surrounding the local hero Rodolfo el Chango Casanova, who won fame in the boxing ring (1970: 280). The *barrio* spirit that Casanova was believed to have defended becomes a major support in Monsiváis's characterization of Tepito as a 'unrivalled synthesis of our communities before they were massified' (ibid.: 288).

From its inception, the image of violence and lawlessness has mostly attached to the markets in Tepito and, above all, the ubiquitous street vendors, who engage in an activity which for the most part is illegal. As John Cross says in his study of Mexico City's *vendedores ambulantes*:

Street vending is a part of the traditional world that intrudes incongruously into the modern. Certainly, it fits our image of the pre-modern world that contrasts with our equally strong vision of modern retailing in strip malls, super department stores and nationwide discount emporiums.

(1998: 84)

In another study, the same author explains some of the reasons why street vending is seen as problematic by policy-makers:

The reason for this is that street vendors are often seen as occupying public space that urban planners prefer to use for other purposes. In addition, the very visibility of street vendors makes them a lightning rod for complaints from established commercial businesses that see them as unfair competitors because they don't pay commercial rents and may avoid taxes and other regulatory codes. In an era of heightened concern for the foreign revenue that tourists can bring, street vendors are often seen as unsightly 'eyesores' to be removed or to be 'packaged' for the tourists . . . The result is that street vendors are often banned or hyper-regulated to such an extent that they have to survive outside the formal, regulated sphere in order to survive at all.

(2000: 1)

The crucial issue, as we can see, is that street vending makes use of public space in ways that are not predicted by urban planners, nor indeed are adequate to the elite's image of the city as a tourist destination, a home for established 'modern' commerce, or thoroughfare. Theirs could be described, following Rogério Proença Leite, as a 'counter-use' of the city.[2]

The degree to which street vending is in fact connected to criminal activities in Tepito remains unclear. One point of contact is *fayuca*, a key concept denoting articles illegally imported.[3] The term likewise denotes the production of pirate products, above all music, dvds, and brand-name clothing. Lonewolf, for instance, claims that Tepito is 'the enemy number one to the music recording industry, as well as to the top-brand name clothing manufacturers' (Lonewolf). The press customarily claims direct links between these illegal activities and other, heavier crimes such as drug trafficking, assassinations, and even a satanic cult (Sánchez 2005). References to a 'Tepito drug cartel' have been frequent in the news, though its existence, apparently, has never been proved. As Piccato says, 'Although neighbours have fought to maintain security and other services in order to preserve a viable community, it would be naive to say that Tepito is not a territory of illegality' (2005). On the other hand, it would also be naïve not to recognize that Tepito, being so near the centre of the city, offers a valuable prospect to speculators who are only too keen to dislodge the current occupiers and release the land, a prime site, for development, being therefore willing to use all kinds of arguments against them.

For their part, the vendors and inhabitants of Tepito are clear about the exaggerations in the press, and their underlying motives. In the following interview, speaking as one of them, the writer Armando Ramírez declared:

This product, it's a lie to say it's *fayuca* or stolen. No doubt some stuff is, but almost all of it comes from clandestine workshops in Iztapalapa, Neza, Tepito and elsewhere. It's really stupid they can't recognize the great enterprenurial talent and the huge creativity that street vending generates. They should channel this and give it what it needs to become legal.

(Rojas)

In 1974, Ramírez and other local inhabitants of Tepito formed the group 'Tepito Arte Acá' (Tepito art here), both to analyze the history of the *barrio* and to try and change it. Through workshops, plays, mural paintings, literature, and local publications, they attempted to appropriate the history of marginality and transform it into their own, positive identity. At the same time, by offering courses and workshops for young people, they showed them other ways of spending their time and opened up the prospect of leaving social marginality behind.

Not long before, at the age of 19, Armando Ramírez had published his first novel *Chin Chin el teporocho* (1972), a picaresque tale, humorously sentimental, of a neighbourhood lad who ended up as drunkard. It is written in a language sensitive to the dialect spoken in Tepito, that depicts life for young people living in a run-down *vecindad*. The *barrio* that emerges from it is in fact a small, fairly quaint place where most people know and deal with each other humorously, undiminished by the general condition of urban poverty.

Ramírez went on to write works which elaborate an idea of neighbourhood that is more complex on several counts, and more problematic than that found in his first novels. In the process, Tepito grows in size, insofar as the setting comes to encompass the neighbouring areas of La Merced and the Centro Histórico. In defining Tepito's world humanly rather than geographically, Ramírez himself now says it occupies the space 'from Templo Mayor and the Palacio Nacional, to the Glorieta de la Cultura in Merced and Avenida Guerrero' (Rojas), an area significantly larger than 'Tepito' is normally understood to be. What links this broader idea of Tepito to the Tepito as it is usually known and as he had portrayed it is the presence of the poor population common to both, likewise with indigenous roots, that lives near or in the Centro Histórico (i.e. at the very heart of what has remained the capital since Aztec times), and their means of survival: street vending. This larger Tepito is also firmly anchored in history through the opening reference to the Templo Mayor, that is, the pyramid and Great Temple of Tenochtitlan, fully revealed to the public gaze only in recent decades, thanks to the extensive excavations that began more or less at the time Ramírez started writing.

Ramírez's works create a universe in which characters from one novel frequently reappear in others, and events from a particular text are re-cast or seen from a different perspective in another. Chin Chin, the protagonist of his first novel, is seen again for instance in a cameo scene in *Sóstenes San Jasmeo* (1997). A proper name of Christian origin, the title alludes to a common item of street-sold merchandise, the brassiere. Chata Aguayo, protagonist of the first-person narrative *Me llaman la Chata Aguayo* (1994), is mentioned in *Sóstenes San Jasmeo* and *La casa de los ajolotes* (2000). The 'house of the axolotl' invoked in this title refers to Tenochtitlan: the Aztec codices tell us that this remarkable amphibian fostered and witnessed the founding of the city (Vollmer 1981; Brotherston 1997: 57). Sóstenes, conversely, appears in *Chata Aguayo*, while *La casa de los ajolotes* may be considered a re-write of *Sóstenes San Jasmeo*, with basically the same plot and the same characters cast in a slightly different way. Both these novels will make references to yet other characters, like the debutante protagonist of *La quinceañera* (1985) and her smuggler-mother, who appear together in an episode of both *Sóstenes* and *Casa de los ajolotes*, and so on.

As a cast or dramatis personae, these characters work together to create a literary space that is intimately connected to the *barrio* of Tepito, augmented now to include the adjoining neighbourhoods, especially the Centro Histórico. Fictional characters rub shoulders with others based on 'real life' people, like actual members of Tepito Arte Acá. The events recounted in the plot allude closely to recent events in the history of Tepito and neighbouring streets that made news in the local press or on TV. Meanwhile, the word 'acá' (here), the trademark of Tepito Arte Acá, remains a constant marker of identity for those who live 'here', that is the enlarged Tepito, as opposed to 'there', that is, elsewhere in the city, especially the richer *colonias* further to the south. All important for understanding them as products and readings of the megalopolis, the geography of Ramírez's novels is clearly determined and relies heavily on this opposition between 'here' in a Tepito synonymous with the ancient heart of the capital and 'there' best exemplified by more fashionable neighbourhoods south of the centre, like Condesa, Roma, and Coyoacán.

On this basis, Ramírez's writing generates a complex concept of 'neighbourhood', one not defined or confined by pre-established official limits or bound by reified bonds between family, custom and tradition but animated rather by fluid boundaries and networks that may reach out into the global economic system. At the same time, in practice almost all the action he narrates happens in Tepito and around Centro Histórico, with the characters hardly venturing beyond the Palacio de Bellas Artes and Torre Latinoamericana, just west of the old centre. This means that they rarely go to the south of the city; in fact, reading some of his novels one might be excused for not realizing that other parts of Mexico City exist at all. A key exception which indeed proves the rule is the novel *Violación en Polanco*, whose implications are distinctly disturbing. It tells of the kidnapping and horrendous sexual torture and murder of a woman from the southern *colonia* named in the title, famous for its central European immigrants and huge wealth.

In the 1980 Grijalbo edition, the author begins the novel by making an extremely strong declaration in the first person plural: 'somos producto de la revolución . . .'[4] As we, he immerses himself in the people who, first, have lived longest and suffered most in Mexico and, second, have been most active in inventing the city's street economy. In both roles their geography finds a common core in Tepito and its market, here explicitly identified by its popular name La Lagunilla. There follows an epigraph from friar Bernardino de Sahagún's sixteenth-century *Historia general de los cosas de la Nueva España*, by implication bitterly sarcastic in its reference to the refined speech (*primores de la lengua*) and moral virtues of 'la gente mexicana'; and another from Céline, rather more straightforward, about time and death. In the story, which cross-references, on the one hand, with Hollywood morality (especially violence and sex) and on the other with exquisite Nahuatl poetry, a group of young misfits (the film of this title is alluded to) seize a woman from Polanco, drive through the whole city in a bus from its southwestern corner, and kill her in the northern outskirts on the banks of the Gran Canal (further filmic references are Sharon Tate and kidnap on a British bus).

Appalling to witness for any reader, the physical details of the torture are constantly integrated into the drive through the megalopolis through allusions, often cryptic, to its history. Too many even to begin to recite here, those cited initially in 'somos producto' include the water-washed stone ('Lajas') of Iztapalapa, seen in the *tlacuilolli* toponym of this place's 1581 *Relación geográfica*, now a distinctly parched and poorer municipality (which houses the local campus of Mexico's Autonomous University, UAM, probably the most important centre of urban studies in Mexico, where García Canclini is a prominent researcher); the nearby Cerro de la Estrella, where the Aztecs observed the 'star' (*estrella*) of its name, the Pleiades, when precisely at midnight they would kindle New Fire and renew time, and which we see penitents still approaching, flagellating themselves in memory not just of Christ's suffering, but Cuauhtemoc's at the hands of Cortés, desperate for more and more gold, and that of the hundreds massacred by Alvarado in Toxcatl (May 1520); and the sprawling industrial slum of substandard housing named after the poet king Nezahualcoyotl (1402–1472) whose verse is ironically cited at length; Tlatelolco, where they outwit the traffic police and where the people, it is said, always resisted 'Aztec imperialism' and in stoning Moctezuma and rallying behind Cuauhtemoc showed the Spaniards they could never win; the *tolvaneras* or dust-storms that afflict the city thanks initially to the invaders' decision to drain the lake. The water was made to flow out, increasingly foul and putrid over the years through the Gran Canal, on whose banks the woman is sacrificed under stars painted 'like vomit' across the sky. This is the gut of the megalopolis or, following the text, the 'monstrópolis', that 'sleeps like a pyramid under a church'.

As the novel progresses, we find out that the crime is actually a revenge, in a plot in some senses comparable with that of *Callejón de los milagros*. It turns out that a girl, one of their group, had been exploited and sexually abused by none other than the husband of their victim, who rich, corrupt and ambitious became by implication the cause of her death and of his own. Sadistic and implacable, the gang are avengers, in the eagle and jaguar guise of Aztec knights. This is perhaps the most unsettling text in Ramírez's repertoire, for it clearly establishes a territorial war, deeply rooted in time, between poor and rich, dispossessed and privileged, old centre and new south, that involves class and ethnicity. The killing of the woman is a sacrifice significant in time and space, in which the poorer part of the city revenges itself by ritually killing its far wealthier counterpart. Ramírez himself sees this novel as prophetic of the violence that at the time of writing was still to overcome the entire city (Rojas n.p.).

MARCHANTA AT THE TIANGUIS

Perhaps the neatest example of how Ramírez's later production defines the idea of neighbourhood is *Me llaman la Chata Aguayo* (1994). The novel is framed as a long video interview given by the protagonist, Chata Aguayo, to her daughter, a student of journalism. The name Tepito is hardly used in the novel: Chata chooses rather to locate her origins by referring to the street, specifically Calle Soledad, which, as she puts it, is far from 'lonely' since it is a 'street made up of six streets' (ibid.: 166). What gives the reader confidence in identifying Tepito as her neighbourhood is, first of all, the fact that Calle Soledad is part of the broad definition of Tepito given by Ramírez (see above). Rather than an official address, this plural street emerges as a rich compilation of resident lives. Also, the term 'acá' (here), trademark of Tepito Arte Acá, is often used to refer to the specific locality where Chata is from and where she works. Moreover, some of the artists and events inspired by Arte Acá are featured in the book as local. Above all, Chata's place is defined by street vending, which, as we saw and shall continue to see, is a (if not the) hallmark of life in Tepito. She is the archetypal 'marchanta' (saleswoman) operating in the very heart of the city.

At the period in her life when she consents to give the interview, Chata is a leader of an association of street vendors in Calle Soledad, but she knows that, because of changes in the political scene, her days in command are almost over. Her tone is exasperated, nervous, and at the same time amused. As in other works by Ramírez, the language is a lively recreation of Tepito popular jargon. We learn that Chata grew up in a run-down *vecindad* in the barrio, only daughter of indigenous parents who spent all their time imbibing *pulque*, the fermented juice of the maguey which of itself powerfully signals indigeneity throughout history in Mexico. The great uprising or 'Corn Riot' of 1691–92 was said to have been plotted in *pulquerías*, establishments that Spanish officials endlessly sought to neutralize and rid the city of (the celebrated swing doors of the Wild West saloon are historically a remote consequence of their efforts). Though often drunk thanks to pulque, her parents still seemed happy and admirable to Chata, because they loved each other: 'I didn't like the *pulque* smell but I liked a lot seeing them embracing, sweating, smelling each other, snore and snore, like two ant-eaters [sexy creatures in folklore], their legs intertwined, embracing, their heads touching' (ibid.: 31).

The whole novel takes place in the run-down streets near the centre of Mexico, in and around Tepito. It is, in that sense, a novel about Tepito, about the activity, the market and street vending, that nowadays perhaps most defines Tepito, certainly in the imagination of the city's mainstream *chilango* citizens, and of people from outside. This activity enables a definition of neighbourhood that is simultaneously local and tied to the kind of national and globalized financial practice exemplified in what we go on to learn about Chata's experience of life.

Cross describes the central areas near the government buildings as the focus of a decades-long conflict between the vendors, who insist on occupying the streets, and the central government, which is pressurized by the middle classes and by the shop owners to eliminate them. According to him, the market or *tianguis* (the modern form of the Nahuatl *tianquiz*), that is, the labyrinthine rows of stands that spread along the streets of Tepito and nearby central areas, serve for a lot of the vendors not just as work place, but also as a home, shielded from sun and rain by roofs of brightly coloured plastic sheets. Many vendors or their workers sleep there at night (1994: 56). Children come with their parents to the vending places, parents cook or eat there, and the majority of the vendors live around the area where they work. In other words, street vending in these areas of Mexico is characterized by the fluidity between private and public life.

Chata Aguayo takes this fluidity to an extreme, not only because she is a woman who works and spends most of her day on the streets (like other vendors, she used to take her small sons with her to the streets, that is, part of her private life was spent on the streets), but also because she becomes a public figure as a leader. At the same time, her private life serves as example and point of discussion in the neighbourhood and among her associates. The framing of the novel corresponds in structure to this blending of public and private. Chata is being interviewed by her daughter both as a public figure (she often advises her daughter about what public use she should make of the tapes) and as a mother who promises one day to reveal to the daughter who her real father is. As she says to her daughter: 'this [film] should be a witnessing of what I am as a leader and as a woman' (ibid.: 106).

Chata's public life is inseparable from her private life in other ways. Sufragio Efectivo, her life-long lover and father of two of her four children, doubles as an important player in her career as a leader. Moreover, his preposterous name, Sufragio Efectivo, which means something like 'voting system that works', comically allows in Dickensian fashion for an allegorical reading of the novel. In it, Chata's love affair with an idea of democracy never brings her a lasting relationship. Sufragio is an attractive, lovable man but also a liar, a cheat, and completely unreliable. Her most trustworthy partner, and father to her daughter, Chimuelo Panzón, is a member of the PRI (Partido Revolucionario Institucional, dubbed 'pirrin' in the novel) who starts his career as a corrupt inspector of street vending, and later climbs the social ladder to become a successful manager of smuggling operations himself, and an assistant to big-time PRI politicians. The disgust her daughter expresses for this father of hers is dealt with sarcastically by Chata. She reminds her that it was he, after all, who had given her the filming equipment, and it was thanks to him, too, that Chata had been able to allow her daughter to study. In other words, for Chata, nobody, at least not the poor, can afford to turn their backs on official power, however corrupt, or refuse to collaborate as a client. Even those who stood in opposition to the PRI are seen accepting its favours.

Chata is a remarkable character, unusual, yet not unlike some of the women leaders in the real-life street vendors associations in Mexico described briefly by Cross (1998: 134–42). As a character who in an interview or conversation gradually reveals beliefs that upset easy western categorization she finds a counterpart in Latin American literature oddly far from home. For in her ways of talking and thinking she closely resembles the principal speaker in one of the great novels of Brazil, João Guimarães Rosa's *Grande Sertão Veredas* (1956), Diadorim. Like Diadorim, Chata is giving an interview in which we never see or hear the voice of the interviewer. We can only deduce her/his comments by the replies given respectively by Chata and Diadorim. Both novels are narrated, as it were, to the 'second person' – that is they are monologues directed at a particular listener. And if Diadorim makes a pact with the devil which allows him to become the bravest *cangaceiro*, that is, a bandit in the socially complex system of *cangaço*, Chata 'sells her soul' to the party in a relationship that is hardly less contradictory and complex and that can no more be simply hailed as resistance than it can be dismissed as reactionary, conservative or (as some would say) 'anti-progressive'.

In turn, Guimarães Rosa has been tellingly compared to his near contemporary the Mexican Juan Rulfo (in a famous commentary on the Latin American novel included by José María Arguedas in his novel *El zorro de arriba*). Like Chata, the characters created by Guimarães and Rulfo are remarkable for the manner and the degree to which they reveal themselves and their situations through speech. In so doing (according to Arguedas 1971), they implicitly appeal to ways of seeing and understanding the world, be it urban or rural, that in going against received western wisdom may be traced back to the prior order of cultural origin openly espoused by Chata and the inhabitants of Tepito.

Chata's idea of *calle* (street) is central to the novel. As we saw, Calle Soledad, for her, is not one street, but six streets, in the first place the ones over which she has authority as an association leader. But *calle* means also the general place where she lives and works, and, above all, it means the *tianguis* and commerce. For that reason, in her view, the street as such did not exist before she started to sell on it and was followed by other street vendors:

> my daughter, look at the street, look at it, look, see, show them the street [with the camera], let them see it, let them see it, how it swarms, how it swarms with people, people who buy, who buy here and there, this and that, who come and go, how much this much and then walking, walking, and I invented it, I, Chata Aguayo, now there are those who say that they were there before, but no way, I am the one, the chosen one, the one who sat on this stool and blessed the street.
>
> (1994: 46)

In other words, she sees the street not as an empty space to be filled with cars or people, but as the people themselves, as commerce, as what happens in the streets as a result of human activity. Hers is a mobile, dynamic idea of the street, of the kind manifest already in Aztec times in a remarkable set of complex terms rooted in the Nahuatl word *tianquiz* meaning to walk around the market or *feria*, to buy and sell, trade or 'mercadear' (*tianquiz-tlayaualoa, tianquiz-manaloyan, tia-miquiztli*), even the phenomenon of sales tax (*tianquiz-tequitl*). For her, all this interchange, cultural, social, economic, is responsible for producing space, in a Lefebvrian sense. It is also, of course, the definition that best suits her role as leader of the street vendors.

Private life and public life form no less a part of Chata's definition of neighbourhood. The 'street' equals vending as a local economic activity, but also as a private activity, that is, a result of private-life conditions which Chata is permanently aware of. When, at the end of the novel, a city official tells her the vendors have to stay off the streets and work in the few market stalls that have been made available to them, she brings in the crucial issue of unemployment. This issue is seen by most analysts as the main cause for the recent massive upsurge of street vending in several Latin American cities and other poor urban areas in the world:

> ['Doctor Gachinski' – the city official]: – There's no other way . . . Now you people will have to start working.
> [Chata] – Working on what . . .
> – There'll be jobs, foreign companies will come in . . .
> – When . . .
> – Once they see we Mexicans know what work is, you people have to show them that.
> – I doubt they're going to pay us what we make on the street.
> – No, of course not, that's impossible, you people are not going to be living in paradise forever.
> – So how are we going to eat?
> – Some in the markets, and others will work in the factories, you have to make the country grow.
> – And me?
> – No, first the country, the future, think of your children.
> – And die of hunger . . .
>
> (ibid.: 172)

Scattered among Chata's generally upbeat discourse about her life there are several poignant moments that allow more intimate insights into her poverty. One touches on her physical and emotional responses to certain tastes, and hence to concepts deeply rooted in her psyche, as it were at a pre-rational level. Such access to the inner world of a character or person is of course well established in (post-Dickensian) western fiction, thanks to the work of, say, Joyce, or Proust. Ramírez recalls how he as a writer was once advised to reflect on the latter's quest for 'lost time' and its archeology, different as social milieu in Proust – sheltered, pampered, precious, hypersensitive –

evidently was from his. The passage in question, which develops La Chata's childhood memories of eating, could in fact arguably be read as a pointed political parody of the famous *madeleine* episode in Proust. There, the taste of a finely flavoured *madeleine* biscuit brings to the adult narrator the involuntary sensations of his comfortable childhood. In *Me llaman la Chata Aguayo*'s generally degraded discourse, the memory is sparked by nothing more sophisticated than Nescafé and bread, and the physical sensation it spurs her to recount is that of childhood malnutrition:

> When I wet a bit of bread in my Nescafé, the memories come to me like amoebas biting my guts, they remove the nausea of malnutrition, I breathe in the smell of Nescafé, the bit of steam moistens my nose, get closer, get closer daughter, do a – how do you say it – a close-up, film just my nose, here, do you see this spot? It was an ulcer [jiote] I had and never went away, from the time I suffered from hunger and used to vomit, when I feel the bread on my tongue, when the crumb gets wet on my tongue, the texture has a strong flavor, but I don't think it is the crumb, it is the experience that stayed somewhere in my mind.
>
> (ibid.: 174)

If Chata has climbed the social ladder thanks to her activities as a street vendor, and a leader of the vendors' association, she also knows that her situation and that of her children is insecure and haunted by fear:

> I know what it is to have all your goods taken away from you, everything you had and you are left with nothing and start crying not crying [llora no llora] thinking how you are going to stand up again, and the fear, the fear that makes you want to become invisible along with all your stuff while the police and the inspectors pass by in their vans.
>
> (ibid.: 60)

Yet, Chata's own account of her life never includes the description of herself as a victim. If she is a single mother who received no material help from the father of her three sons, she also describes herself as sexually attracted to him. She is a sexually liberated woman, aware of her physical needs and limitations. Read this way, one could think the novel presents us with a celebration of street vending in the style of the neo-liberal economist Hernando de Soto:[4] Chata would be the prototype of the 'small entrepreneur' in the 'informal economy' who has made it to a more comfortable life. But this reading would not be quite correct. The idea of 'free enterprise' in the novel is balanced by the relationship of dependence that Chata and all her allies are careful to maintain with the representatives of the then official party, the PRI. This relationship is historically validated:

> The vast majority of regular street vendors in Mexico City belong to associations ranging in size from a few dozen to 7,000 members. This is not,

however, simply the result of 'grassroots' organizing. Rather, . . . it is a direct result of administrative procedures requiring individuals to form part of a 'recognized' association before being allowed to sell in the street. Furthermore, most of these associations are affiliated with the PRI.

(Cross 1998: 120)[5]

One of the most frequent words used in the novel, and in Ramírez's works in general, is *transa*, that is, corruption, illegal dealings. Chata is *transa*: she makes legal and illegal deals with city officials and pays a large part of all her earnings to Chimuelo Panzón, her partner and the party representative, so that she and her associates can go on selling on the streets. The relationship is usually described by political analysts as co-optation or clientelism 'that serves the interests of the Mexican state, supporting the idea generally accepted among such authors that the Mexican state most strongly reflects a "strong polity" model' (ibid.: 120). Cross, however, argues that the relationship as such is more complex:

> However, this policy also gives the leader substantial power not only over their own members – vendors who do not submit to their authority can be ejected from the market – but also relative to the same officials who create them. That is, the leaders can use their membership base politically to outmaneuver low-level officials charged with controlling them and to appeal to clientelistic relations that undercut the administrative power of the city.
>
> (ibid.: 122)

Me llaman la Chata Aguayo well illustrates that point at a general political level, yet as literature from within it may also be said to go deeper, emphasizing Chata's simultaneous strengths and weaknesses and also at a personal level. Her discourse is set against the protests of her daughter, whom we never hear but whose comments can be deduced by Chata's own responses. As a university student who hates the PRI and its reputation for corruption, Chata's daughter is a mirror of the novel's most likely reader in the condemnation of *transa*. Chata, however, simply sees *transa* as part of life, as the only way up for poor people like her.

The novel itself does not espouse *transa*. Being structured as a first-person narrative, it allows us to glimpse Chata's contradiction and anxieties. Many of her characteristics are meant to make her ridiculous to most readers, thereby adding humour to the novel. She gets unfamiliar words wrong (also a Dickensian ploy) and for instance says *antropófago*, cannibal, when she means *antropólogo*, anthropologist), her hair is dyed an improbable blond, she worships the president as if he were a god, and so on. Yet, Chata is a likeable character, admirable for her courage and liberation. Her capacity to change, to adapt to new circumstances and adopt new ways of living makes her the opposite of a bastion of traditional *barrio* custom, even though she is depicted as a typical representative of Tepito and Centro Histórico culture, the culture of vending. If she is corrupt, so is the rest of the society she is immersed in, above all here right at the geographical centre of national power. Commenting how a luxury shopping centre was built on what was supposed to be an ecological reserve (as Walmart, in fact, was at Teotihuacan), she compares the morality of the deal to her own *transa*:

> How do you think they built this shopping centre, you see, it is an ecological reserve, it is not likely the health officials will come and talk to the manager to see why it smells like popcorn and butter. Who doesn't see that so much glass is dangerous, all these escalators, all these elevators, what about their permits, no, they gave them a chance, no inspector will mess with them, they will go to the President.
>
> (Ramírez 1994: 207)

This is probably the novel's strongest achievement: the way it presents the protagonist as an ideologically complicated subject, at once an icon of resistance in the occupation of urban space and a co-opted and corrupt PRI affiliate, with ambitions of becoming a congresswoman. In one person, she is both an Indian girl, capable of turning her ancestry to political advantage, and a false blonde consumer of Coca-Cola and hamburgers who in no circumstances would want to be painted by Diego Rivera as a typical Mexican Indian woman. In the same breath she defends both the poor, who come to the streets of Centro Histórico to sell and buy, and rich corrupt politicians. Characters of such complexity, products in their way of the megalopolis, are not easy to find in Dickens, whose work was after all much relished and promoted by nineteenth-century capitalists.

Through Chata's contradictions, the novel creates a sense of place anchored on a complex notion of street vending which refutes the image of it that is promoted in the press and the media. As Cross puts it: 'Street vending has been criticized for causing or contributing to a number of social ills that afflict Mexico City. Indeed, rhetorical claims by critics of the practice often blame it for being the source of whatever is the latest problem city officials have identified' (1998: 109). The novel makes the same point in a satirical vein, through the city official that tells the protagonist:

> you made the toy industry break, you made the big stores go bust, it's your fault that the country is left without its architectural heritage, it is your fault we're not doing well in history, it is your fault that IBM, Kakanasaki, Fud Chemical Easternoon get scared away, they look at you and go back to their countries, no, the miracle is over now: you have to go back to work.
>
> (Ramírez 1994: 172)

Against the argument that the street vendors destroy and mess up views of historical buildings and monuments, Chata says: 'It is not like before, when people didn't care whether the Centro Histórico was the heritage of mankind [patrimonio cultural de la humanidad], of course screw the humans who live in those places' (ibid.: 253).

Street vending then becomes the central activity in a movement not only to occupy the space of the city, but to recover it. Chata describes it as an activity already practised by the city's ancestors: 'the street vendors will not disappear, because we've been coming here since the days of the great Tenochtitlan, that's the way we imagined selling and that's the way we have managed to get others to buy our stuff' (ibid.: 206). And it is not only vending itself that defines their occupation of space, but also consuming, buying: 'Before, just commerce got money from the rich and the middle class, nothing from the poor, and we sell to the down-and-outs [los jodidos], here, as in the old *tianguis*, like when the Aztecs went: Hey, saleswoman, how much?'. (ibid.: 124).

Nestor García Canclini has developed his idea that citizenship in the contemporary world is being defined more and more by patterns of consumption, in the following way:

> For many men and women, especially youth, the questions specific to citizenship, such as how we inform ourselves and who represents our interests, are answered more often than not through private consumption of commodities and media offerings, more than through participation in discredited political organizations.
>
> (2001: 5)

Consuming, according to him, is not only 'a mere setting for useless expenditures and irrational impulses' but also a 'site that is good for thinking, where a good part of economic, sociopolitical, and psychological rationality is organized in all societies' (ibid.: 5).

Much amplifying Marx's definition of 'market' in its own terms *Me llaman la Chata Aguayo* defines neighbourhood as part of a territorial dispute that is concerned not only with the right to sell but with the right to consume. In this light, it is worth noting that most discussions about street vending focus exclusively on the sellers and their occupation of territory, leaving out the thousands of people who visit those market places on a daily basis. After all, street vending would not work were people not willing to buy and consume. If consumption is a manifestation of citizenship, as García Canclini postulates, then we should start studying the patterns and movements that guide the desires and actions of the consumers who buy at street stalls in cities all over Latin America, the same stalls that are execrated by the media and much of the middle class population. Going back to Chata, her definition of place, or *calle*, includes both salespeople and consumers: 'they should see the rivers of people that come down from the North and the East of the City, from the poor places, how they arrive from the mountains, from the lost cities, from Chalco, Neza, Villita mountain, taking everything back with them from calle Soledad' (Ramírez 1994: 254). That is, the poor people from the outskirts and heavily indigenous areas of the city are look-ing to exert their rights to the same things that richer people in the city consume, to be equal citizens, at least in this rudimentary and pragmatic definition, of the megalopolis.

In Chata's case, selling and consumerism can never be divorced from the phenomenon of *fayuca*, commodities pirated or sold on the black market, illegal yet of a complexity allowed in turn to emerge in the novel and elsewhere in Ramírez's work. On the one hand, *fayuqueros* are ridiculed in texts like *Quinceñera*, the mother of the protagonist having made her pile thanks to the *fayuca* operations. On the other hand, *fayuca* may be represented as just a mode and means of commerce, as honest or dishonest as any other. As Chata says of herself: 'I started like Mr. McDonald's' (Ramírez 1994: 123). When she has problems with the police because of selling items with forged brand-names and pirated music, she compares herself to the Koreans and Chinese, implying that the success of their commercial activities is based, as is often alleged in the West, on 'copying' Western products. Besides being a street vendor, Chata and her son runs a recording company that produces pirate CDs. She argues she uses technology to produce music that sounds better than the original, to enhance rather than simply copy it (ibid.: 157). Through the sale of pirated items and falsified goods, Chata and other street vendors see themselves as allowing the poor to have access to the consumption of goods that look like and sound like (or somewhat like) the goods sold in expensive malls. In this logic, *fayuca* and pirated products are, on the one hand, a by-product of the fetishism of brand names in contemporary consumer markets. On the other hand, they are an appropriation, by the poor, of the contemporary logic of simulacra. Discussing the issue of pirating with regard to the record industry, Ana María Ochoa reminds us that it is far from being a good-versus-evil proposition:

> But as Jack Bishop observes, this [statistical] language must be interpreted and reformulated in the light of a complex net of relationships between types of profits by the industry (for instance some of the giant com-panies make money not only with the records themselves but also with the blank cds, that is, with pirating), managerial salaries, profit with each record, practices like price fixing at a global scale, no matter what the local salaries actually are in the different countries, quota policies for membership in transnational music associations (which cannot be afforded by local independent companies) and so on.
>
> (2003: 79)

The themes of pirating and *fayuca* are carried forward in Ramírez's next novel. *¡Pantaletas!* (2001, Panties!), as its title forewarns, is a generally lighter-hearted take on street commerce in the Centro Histórico and greater Tepito. True to his Dickensian penchant for giving his characters funny and obviously allegorical names, the author calls the protagonist of this story

Maciosare. This is to be heard as a phrase in Mexico's national anthem 'más si osare' (if they dare), misunderstood and misspelled. The novel is a sarcastic analysis of how the dream of modernity fails poor people of indigenous origins. The protagonist's parents believed in PRI propaganda about modernity: his mother practised birth control and had only two children, and believed in education above everything, so that, at enormous cost to herself, she sent her son to university. Maciosare studied sociology there, but unable to get a job afterwards that would allow him the means of survival, he ended up becoming a street vendor. After a tough start selling tacos for his father-in-law, he became mildly more successful selling women's underpants in larger sizes. A satire of both the positive discourse on micro-enterprising (which, in line with de Soto, tends to laud the creativity of the 'informal economy') and the attacks made by the press on the Centro Histórico street vendors, the novel also deals with the subject of piracy and smuggling. Maciosare's mother, for instance, is harassed by the police for selling Mickey Mouse socks, and her arrest becomes sensational news on television

> My mother was in the news on television tonight. She was identified as a threat to the country's sound finances; her activities sapped the confidence of financial capitalists on Wall Street.
>
> George Soros should be careful investing around here and all because of my mum and her Mickey Mouse socks.
>
> Business leaders accused her of not paying tax. Academics warned her that if we wanted to be a first-world country we'd have to respect the law and punish offenders. Party politicians with power portrayed her as an 'emissary from the past'.
>
> . . . pirating, tax evasion, selling stolen goods, interfering with rights of way, damage to the nation's cultural heritage . . . along with offences against the environment and destruction of historical monuments.
>
> (ibid.: 112–13)

Loaded with sarcasm, this account rehearses many of the arguments used by the press to attack street vending. Against the argument that it destroys the historical monuments, Maciosare's answer is stronger still than that of his predecessor, Chata Aguayo: 'they argued that we were destroying their art treasures, as if every stone of these buildings did not belong to us and was not soaked in our drops of blood, sweat and tears as it was with our grandfathers' and will be with our children' (ibid.: 143).

Armando Ramírez's novels emphasize that the arguments over street vending try to disguise a war for the space of the city and its uses. It is less about vending as such than about who has the right to occupy the streets. His arguments become even clearer when set against, for instance, the *crónicas* of José Joaquín Blanco, who writes regularly in periodicals like *La Jornada*, *El Universal*, and *Nexos*. In a *crónica* written with Jorge Olivera

Ramos, he defines la Merced (where Chata's calle Soledad is located) as an *antibarrio* created by the eighteenth-century liberals, who wanted markets to return to the centre of the city after the Bourbons had sent them packing to the outskirts. This is how he describes the Bourbon attempts to get rid of the central markets: they wanted

> [to] place out on the *periferia* the stomach, genitals, and excrement, the most basic daily needs and routines of society, so that the prestigious part of the city could be filled just with public and business offices, as before it had been smothered in churches and convents. A peripheral *tianguis* city.
>
> (Blanco 2003: 53)

Blanco and Olivera Ramos recognize that the attempt did not work completely, that some vendors: 'always stayed in the centre, a majority, but hated and harassed, promoting themselves as a thick almost shameful mass' (ibid.: 53). But worse still, in their view, was the liberal's desire later to bring the markets back to the centre on the grounds of convenience. The result, for the authors, became: 'Well, neither a pretty model city nor markets worthy of the name, but ruins and a chaotic mix-up of both' (ibid.: 54) ['Bueno: ni bonita urbe de maqueta ni mercados dignos, sino ruinas y revoltijo caótico de ambos']. The centre has now become, in their view, a market by vocation:

> mansions and even apartments converted into warehouses and factory outlets; lavatories (from 50 centavos to 2.5 pesos at the end of the 20th century), public baths; streets nothing but loading and unloading, overflowing with bloated trash, and alongside one of the cheapest and most horrific brothels in the world, thriving and well stocked with rats and flies; decrepit churches that also have a certain brothel-like air. Indians, prostitutes, bums or *tamemes* needy as beggars, covered in scabs and bugs, savage and insulting.
>
> (ibid.: 54)

Witty and informed as Blanco and Olivera Ramos undoubtedly are or are found to be by their readers, their dyspeptic assault on the poor leaves a bad taste, and ultimately has racist implications. The *crónica* does not attempt to hide that for the authors, the space of La Merced, that is, the centre of the city, should, in their view, be reserved not only for certain functions of the human body but also for just certain classes of bodies. For, although street commerce in the centre concentrates on thousands of products of various kinds (in another *crónica* Blanco refers disparagingly to those who 'sell anything'), here they re-define the market by talking only about Indians, prostitution, down-and-outs, beggars, and *tamemes* (in pre-Conquest and colonial times an Indian who carried goods), that is, again definitely Indians. In another *crónica*, Blanco complains about how the Paseo de la

Reforma, the magnificent avenue that runs south-west from the centre to Chapultepec, has been invaded by 'tribes of indigents like swarms of gutter Apaches'. Juxtaposed as it is by terms like 'tribes' and 'apaches', the word 'indigent' (*indigente*) inevitably becomes associated with 'indigenous' (*indígena*), or, in a Spanish pun, 'gente indígena', that is, indigenous people (ibid.: 12). In support of this reading comes yet another *crónica*, in which the same author attributes the beginning of the end of Zócalo, to the restoration work promoted by President José López Pórtillo:

> arrogant López Portillo style archeologists, who destroyed the superb panorama of the National Palace in order to adorn it with a wasteland of stone foundations: the rubbish tip of the Aztec Templo Mayor! They recovered no temple (*teocalli*) for us; they just put on show a crater of rubble, a monumental scar. Long live the story of our nation in rancor and ruins!
>
> (ibid.: 11)

Very much in the opposite camp, Ramírez founds the idea of Tepito on the right that the poor, the underprivileged, and the Indians have to go on occupying it. Their right to this space, at the very centre of Mexico's megalopolis, is defended by the work they do, by their own creation of space, which emulates the space occupied by their direct (or indirect) ancestors. In Blanco's discourse, the poor (i.e. the Indians) are responsible for the deterioration of the city, for being an embarrassment (in whose eyes?), for threatening the modernity of the city by their very presence, through their undesirable bodies and its undesirable parts, and with 'their' ruins. ¡*Pantaletas!* offers a satirical response to views of street vending as a backward, anti-modern activity, by situating it rather at what is now suspected to be, at least in the megalopolis, a kind of avant-garde in the global economy:

> the informal economy is the most democratic road to the free market, to minimum interference from the State and its actions ought to be collaborative: listen, friend Macas, I'll cut off one of my balls and half of the other if the informal economy doesn't take over more than two-thirds of world business. It's the most efficient way of spreading out globalized wealth.
>
> (Ramírez 2001: 121)

Megalopolitan to the core in Mexico City, Ramírez and his *marchanta* negotiate and articulate life in Tepito in terms that will in part be all too familiar from standard social science accounts of the very large city. In telling their stories from the inside, to that extent they augment and offer to modify and refine those accounts. The 'contradictions' that abound in La Chata's discourse are less than when located in her market and seen to grow from the long urban experience its exchange implies, in the street space it generates. Her engagement with the state apparatus, repressive and ideological, is best understood on that basis. She and her fellow vendors resist the police and a terrible press buoyed by belief that they are the ones to have occupied the city longest. If they fail to froth at the notion that the PRI is endemically corrupt, then it will be because that political party, however miserably untrue to its revolutionary mandate, still enabled a *modus vivendi*, not outright eviction but possible transaction in the face of traditional hostility rooted ultimately in the bad conscience of the invader and in racism. In the condition to which she sees herself and her people being reduced, despised and negated, to survive at all becomes a main motive for pride. Prior to the triumphs of the neo-liberal crusaders at the end of the last century, that pride could be expressed through adulation of people from the *barrio* who left and prospered, as a kind of celebrity, for example, the boxer from Tepito so perceptively analysed by Carlos Monsiváis back in 1967. Today it has becomes inseparable from knowing you know the codes of life in the megalopolitan *barrio*.

5 Capão Redondo and the space of rap

During the celebrations in 2004 of the 450th anniversary of the founding of São Paulo, the daily newspaper *Folha de São Paulo* asked citizens to name places in the city that they considered the most beautiful, and the ugliest. Though there was no single 'winner' for the latter category, slums or *favelas*, according to the newspaper, stood out as most persistently voted 'the ugliest'. The *bairro* Capão Redondo, distinct as such from the *favela*, was among the very few in the neighbourhood category to enter this list of undesirables.[1]

The great majority of the people who responded to the poll probably have never been to the remote Capão Redondo, a *bairro* of 200,000 inhabitants on the south side of the city. But they might well have heard its name. In the past ten years the neighbourhood has acquired a reputation in the press for being an extraordinarily violent, drug-ridden place, where life is worth less than a pair of sneakers. Such a reputation is at least in part justified: according to some statistics, the probability of a person dying of violent death in Capão Redondo is 46 times higher than in other parts of the city (http://www.depoisdotiro.blogspot.com.br). Data from the early 1990s put Capão Redondo as the area in São Paulo which had the lowest standard of living, along with the highest rates of illiteracy and infant mortality; and the largest proportion of inhabitants living in slums within it. In the same statistics, Capão also held third place in the city for homicide – the first place being awarded to another part of *Zona Sul*, or the south side, where Capão is located (Maricato 1996: 89).

Situated in the district of Santo Amaro, a former agricultural town that was incorporated into the city of São Paulo only in 1935, Capão Redondo is, unlike Tepito, a new neighbourhood. Santo Amaro was occupied before the European invasion and became the site of a Jesuit school built in 1560 for the purpose of evangelizing the indigenous population, and many of the place names in the surrounding region are of Tupi-Guarani origin. Capão Redondo never formed part of an urban centre, however, again unlike Tepito, indeed it always lay very far from it, on unenviable outskirts that now are generalized in the term *periferia*. Over the centuries the numbers of its original inhabitants were greatly reduced by slavery and massive immigration. Capão Redondo means 'a round area cleared of vegetation', a place-name traceable back to 1912 (www.capao.com.br), which could suggest that it was used for hunting, though hunting what is unclear.

From the 1950s, and above all in the 1960s and 1970s, Capão Redondo and other neighbourhoods in the *periferia* of São Paulo became home to the tide of immigrants from the Brazilian North-east, who were attracted to the region because of the relatively low value of land and the non-existence of planning and building regulations.[2] Capão Redondo is therefore home to people who mostly came from somewhere else, driven by poverty. It is a neighbourhood of displaced people who, because of they are mostly from North-eastern Brazil, suffer strong prejudice from other *paulistanos*.

Neighbourhood literature in São Paulo has always been linked to newly arrived populations. Between 1910 and 1920, Alexandre Marcondes Ribeiro Machado published satirical *crônicas* about the Italian neighbourhood of Bixiga under the pseudonym Juó Bananere, and stretched Portuguese to its limits, in portrayals of immigrant behaviour in the city that are blatantly funny. In the following decade, Patrícia Galvão's novel *Parque Industrial*, and the short stories by Antonio de Alcântara Machado (*Brás, Bixiga e Barra Funda*) also concentrated on the neighbourhoods inhabited by Italian immigrants, especially Brás and Bixiga. These were the first *paulista* neighbourhoods to be perceived as such, to acquire what was thought to be a certain kind of notoriety for having a distinguishable culture. Nowadays, Bexiga is still São Paulo's tourist 'Italian neighbourhood', figuring in guide books as the place to eat Italian food in noisy restaurants, and exulting in an annual *festa* of its own. Brás, on the other hand, became home to new waves of immigration, mostly of Northeasterners, and also had the honour (along with Capão Redondo) of appearing in the *Folha de São Paulo* poll as one of the ugliest places in the city.

Galvão's *Parque Industrial* (1933) bears the subtitle 'a proletarian novel', and describes the lives of women who work for a fabric factory in Brás. As the sub-title forewarns, as 'proletarians' the women are portrayed as exploited workers, and they follow different paths in their lives. The beautiful mulatta Corina is seduced by a man from a richer part of town, while Otávia becomes a revolutionary leader. Brás is depicted in the novel as the quintessential working-class neighbourhood, full of identical-looking houses and run-down *cortiços*, all overshadowed by the massive presence of the nearby factories with their chimneys.

Alcântara Machado's collection of short stories, *Brás, Bexiga e Barra Funda* (1927) had already offered a fuller view of everyday family life of immigrants, many of them children. The emphasis here is on survival and adaptation of immigrants to a city that was changing very fast, and to their forging a new language: an italianized version of Portuguese that still today characterizes, in many senses, the spoken tongue of São Paulo as a whole. Alcântara Machado's more sympathetic view of working-class neighbourhoods would be continued many years later, in the 1960s, in João Antônio's short stories, set mostly in *periferia* neighbourhoods and focused on characters such as billiard players, drunks, and thieves. His stories are quintessentially urban, and tend to concentrate on different parts of the city, on a general portrayal of *periferia* rather than on specific neighbourhoods.

The most ground-breaking contribution to the fact and concept of neighbourhood literature in São Paulo before the 1980s was made by a woman who was never even considered a writer, let alone one who wrote about neighbourhoods. She is Carolina Maria de Jesus, who began her days in a *favela*, living off scrap paper she gathered from all over the city, and the autobiographical account of her trials made her world famous. Her *Quarto de Despejo* (*Child of the Dark*; literally more like 'junk room') caused a scandal when it was first published in 1960. Newspapers doubted that she had been the actual writer, and suspected Audálio Dantas, a journalist who helped her manuscript reach printed form, of having written it for her. After its noisy publication, the diary practically disappeared from Brazilian literary histories, at the same time that it became quasi-obligatory reading in courses of Brazilian culture in Europe and the US. Jesus's *Quarto de Despejo* is an important predecessor to the cultural phenomena that interest us most here: literature and music born in the poorer parts of the city, directly witnessing everyday life in the *favelas*. Jesus's diary ushers us into the *favela* without claiming for it the status of a neighbourhood or *bairro*. For her, the *favela* was a temporary place, a place that was bound to disappear once Brazilian society became more just, or – slightly less improbable – a place that with luck she would leave and move on from once her life improved. In her rather unusual case, this eventually happened, thanks precisely to the publication of her book.

By contrast, since the 1980s in Capão Redondo, a cohort of young men is responsible for a strong cultural production that deals with everyday life in the *favelas* while being, at the same time, definitely centred on an idea of *bairro*. It all started in 1988 with the foundation of the rap group Racionais MCs, two of whose four members, Mano Brown and Ice Blue (both black), are from Capão. Their fourth record, *Sobrevivendo no Inferno* (Surviving in Hell), won several prizes and turned them into the first Brazilian rap group to sell millions of copies. Following on their steps, many other hip-hop groups emerged in Capão and the surrounding neighbourhoods. These new groups tend to organize themselves, like their counterparts in the US, in what are known as *posses*, that is, congregations of various hip-hop groups who meet regularly to discuss and study music, dance, and graffiti techniques, and to do educational work in the community.

In 2000, Ferréz, himself a rapper, published the novel *Capão Pecado*, which describes everyday violence in Capão Redondo, appealing to the idea of its 'sin' (*pecado*) as the crimes and the wrongs that make it a blot on the city. The novel includes as series of 'vignettes' to introduce each chapter, written by many of the most important rappers from the region, among them Mano Brown, from Racionais MCs, and Conceito Moral (Moral Concept). In 2003, Ferréz published a second book, *Manual Prático do Ódio* (A Hands-on Guide to Hatred), and has become the editor of a page of 'marginal literature' published more or less regularly in the leftist monthly *Caros Amigos* (Dear Friends), whose readership extends throughout Brazil, and beyond. His books have also sold in large numbers and his name appears quite frequently in mainstream national dailies like *Folha de São Paulo* and *Jornal do Brasil*. What he calls 'marginal literature', that is, the publication of texts by *favelados*, landless workers, prisoners, semi-literate people, is connected to a recent interest, in Brazil and Spanish America, in such genres as *testimonio* literature, prison diaries, and works written in contexts of violence or from the most disadvantaged sectors of society.[3] In the case of Férrez, literary production goes hand-in-hand with attempts to re-create Capão Redondo as a space of resistance, starting with his own pen name, Férrez: he says it combines the first syllable of Ferreira – the name of the revered *cangaceiro* Lampião (the type of North-easterner who inspired Guimarães Rosa, see p. 93) and the first letter of Zumbi, the leader of the largest *quilombo* (the maroon runaway slave community) in Brazil.

In Capão, the year 2000 also marked a key technological broadening of media accessible to the *bairro*, when the brothers Allan and Leonardo Lopes founded the website www.capao.com.br (henceforth capao.com). They said their aim was to show the 'good side of Capão' (capao.com). Later, the website started to post useful information, and now, in its third phase, it has an open format and invites responses, comments, stories, poems, thoughts, and publicizes bands, events, meetings, open days, help programs, and so forth.

The three phenomena, rap, Ferréz's novels, and the website, are interconnected, and ask to be treated as such. Rap, however, functions as the central focus and the main reason for them all. They may be discussed generally, though not unproblematically, under the rubric 'youth culture'. As Rossana Reguillo Cruz points out, 'youth' as a social category and a target consumer group, was invented only after World War II (2000: 21) and it suffers from the fact that today's youth is tomorrow's old man. Indeed, the term has been criticized by several scholars, among them Bourdieu, who famously commented on its limited conceptual possibilities, since age is perceived differently in different societies and social classes (1984: 143). Rupa Huq, in *Beyond Subculture* (2006), rehearses many of the arguments against the use of youth as a category in order to arrive at a less Eurocentric definition of both 'youth' and 'culture'. This is not the place to engage in lengthy discussion about the rubric, which is useful here only insofar as it helps locate the Capão groups and keep them together, along with their cultural production, as part of a supra-national, global phenomenon. At the same time, the artists to be discussed fit only problematically into the category, not least because most of them have grown out of the age in which they could easily be classified as 'youth' (they are now mostly in their thirties) and have become public figures in ways that go beyond their 'youth culture' origins.

Most definitions of hip-hop explain that it is a broad phenomenon that includes rap music, break dancing, and graffiti, and São Paulo hip-hop is no exception. Histories of *paulista* hip-hop trace it back to 1983, when break dancing started to be practised in certain dance clubs in the city. Nelson Triunfo, a recent arrival from the North-east, took the dance to the centre of the city, where a group of young men started to meet regularly to practise, listen to rap music, and organize graffiti expeditions. In 1988, the producer Milton Sales brought together two young men from the south side of the city whom we have already met, Mano Brown and Ice Blue, and two from the North, Edi Rock and the DJ named KL Jay. As a result the group Racionais MCs was formed, soon to become the most successful hip-hop group in Brazil.[4] Before the year was out (1988), they had placed a few songs in an anthology of rap. In 1990, they went to record their first solo LP, *Holocausto Urbano* (Urban Holocaust), followed by *Escolha seu Caminho* (Choose your Way, 1992), *Raio X do Brasil* (X-ray of Brazil, 1993), *Sobrevivendo no Inferno* (Surviving in Hell, 1997), and the double CD *Nada como um dia depois de outro dia* (Nothing like a Day after Another, 2002). The first three releases were produced and distributed by the rap co-operative Zimbabwe. After 1995, Racionais MCs created their own label unreassuringly named Cosa Nostra.

Racionais and other groups from São Paulo are clearly part of a larger global phenomenon which in the last two decades has seen the impressive spread of rap music and hip-hop culture throughout the world (Osumare 2001; Mitchel 2001). Though rap can be heard in other regions of Brazil, São Paulo was the first city in the country to see a massive growth in its popularity, and, along with certain parts of Brasília, the only one where an idea of place has been connected to this particular musical rhythm.[5] Racionais have found inspiration in US hip-hop groups like Public Enemy and 2Pac (Tupac Shakur), and what Tricia Rose calls the 'brilliant chilling stories of ghetto life' by Ice Cube (1994: 295). From these US groups the Brazilians received the general musical conception and beat. They learned, too, the creative use of 'sampling' by the DJs, who, in rap, convert recorded music, turntables, and mixers into musical instruments; and the use of the MCs, Masters of Ceremony, whose job it is to usher in the numbers through talk and singing that become their lyrics. Much of the terminology used in Brazilian rap and hip-hop in general is English: MCs, DJs, B-Boys, scratching, etc., as well as the names of the artists (Brown, Ice Blue, Lady Rap). The group name that Racionais MCs chose, however, blazes its own trail in asserting faith in reason and the need to think, concepts eminently borne out in their lyrics. For reasons of space, break dance or graffiti will receive little attention here, the focus is rather rap, since it ties in most closely to these other two phenomena.

Rap is well known for its strong spatial sense. As Murray Forman explains, noting the intriguing rehabilitation of the term 'hood' (short for 'neighbourhood', the term was also formerly used with the meaning of thug):

> Since rap music's inception, its lyrics have articulated the details of place with ever greater specificity. In the process, MCs have transformed the

abstract notion of space into a more closely defined locus of experience as the close-knit relations that cohere within neighbourhoods and city blocks are granted discursive primacy. If space is a broadly configured dimension, place, as framed within discourses of the 'hood', constitutes a microscale of experience that has, since roughly 1988, achieved greater significance within hip hop; today 'the "hood"' prevails as hip hop's dominant spatial trope.

(2004: 156)

The emergence of rap in New York City in the early 1970s draws strongly on the fact and imaginary of two spaces: the Bronx and Harlem. Rap brings ghetto life into the music, not only because it *talks* about everyday experiences in particular places, but also because it transforms those experiences into music: it is sung in neighbourhood and Black dialect, and embraces a gamut of urban noise: cars, gun shots, telephones, radios, people talking, and so on. As Robin D. G. Kelley observes, rap also interferes in public space:

> music and expressive styles have literally become weapons in a battle over the right to occupy public space. Frequently employing high decibel car stereos and boom boxes, Black youth pump up the volume not only for their own listening pleasure, but also as part of indirect, ad hoc war of position.
>
> (1996: 115)

References to physical space, to particular neighbourhoods (or 'hoods') are integral to rap, and the 1990s feud between the West Coast and the East Coast chapters, which ended in the actual assassination of Tupac Shakur and Notorious B.I.G., had after all a strong spatial dimension.

Racionais MCs' spatial sense derives to a large degree from the US idea of 'hood'. But it is also deeply connected with São Paulo's south side, and more specifically with the neighbourhood of Capão Redondo. Paul Gilroy's view of rap as a diaspora music form may help us look at Brazil's particular development of it as part of a historical process that he describes in the following way:

> the transnational structures which brought the black Atlantic world into being have themselves developed and now articulate their myriad forms into a system of global communications constituted by flows. This fundamental dislocation of black culture is especially important in the recent history of black music which, produced out of the racial slavery which made modern western civilization possible, now dominates its popular cultures.
>
> (1993: 80)[6]

Racionais themselves express a similar view of Black music in the song 'Fio da Navalha' (Razor's Edge), in which they play not rap, but instrumental blues: 'Black music is like a big tree with many branches and so on / Rap is one, reggae is another, and samba too / Now we are going to show another one'. As Murray Forman points out:

A growing body of scholarly research . . . indicates that today hip-hop is also part of the everyday practices and experience of youth around the world as they combine its expressive forms with their own national and local inflections. DJs, MCs, B-boys, and graffiti artists in dispersed global contexts are actively integrating hip-hop with their own cultural experiences, creating new locally relevant meanings and redefining their social environments as they do so.

(2004: 157)

It is also important to note that the social economic conditions that saw the emergence of rap in São Paulo in the mid-1980s and early 1990s bear many similarities with the conditions that gave rise to rap in many US cities. This is particularly true of what Tricia Rose describes as the background conditions that stimulated the appearance of gangsta rap in the West Coast in the late 1980s:

During the late 1980s Los Angeles rappers from Compton and Watts, two areas severely paralyzed by the postindustrial economic redistribution developed a West Coast style of rap that narrates experiences and fantasies specific to life as poor young, black, male subject in Los Angeles.

(1994: 59)

It was also the post-industrial reality of São Paulo, where the chances of a young uneducated Black man getting a job were (and continue to be) minimal, that created the São Paulo rap.

Racionais and other rappers use many terms to refer to their own place, and to connect it to places around it. The most frequent are 'area', 'quebrada', and 'pedaço' – all of which are commonly used words for 'neighbourhood' in an unofficial sense (that is, they are general ways to refer to the 'area where I live'). They also employ the word 'gueto' (ghetto; echoed in the title of their first LP *Holocausto urbano*), and cite such unappealing or dangerous locations as inferno, Vietnam, underworld, and so on. They often refer to Capão Redondo and its various internal divisions by their proper names (Favela do Fundão, Jardim Rosana, Cohab Adventista). Spatial specificity is in fact a hallmark of their songs: we find names of streets and roads, areas within the neighbourhood, apartment complexes, neighbouring *bairros*, rivers, hospitals, schools, etc. There are also more intimate types of toponyms, which can be understood only by locals and best imagined from their psychic centre, like 'back street', 'little field', the 'asphalt road', etc. Such specificity is enhanced by the frequent inclusion of names of people, other rap groups, dialogues, and so on. By referring often to their neighbourhood, to streets and places in it, to people they know locally and that know them, Racionais songs delineate a strong local geography.

At the same time, they (along with Ferréz and the website) locate Capão Redondo as part of Zona Sul, São Paulo's south side, an expression used again and again to denote an underprivileged area of the city. In fact, Zona Sul itself includes many wealthy neighbourhoods. Their use of 'Zona Sul' to refer exclusively to those that are far less so, there and elsewhere in the city, is a strategy of incorporation, of making the space their own. Capão and Zona Sul are also described as part of *periferia*, that is, the poor areas in the outskirts of the city. The use of *periferia* by rappers and many youth groups is so frequent and so insistent that the Worker's Party (PT) congressman Eduardo Greenhalg is reported to have been amazed, after a community meeting, by what he called 'periferia patriotism' (Guasco 2000: 108).

It is notable that this strong division between *periferia* and *centro*, which conceptually corresponds to *periferia* and the richer parts of the city, matches the received account of urban development in São Paulo. It differs, in that sense, from Teresa Pires Caldeira's ground-breaking (2000) study, *City of Walls*, where she argues that the division between *centro* vs *periferia* is being replaced in São Paulo by a closer physical proximity between wealth and poverty, a proximity that has, according to her, generated spatial strategies of separa- tion like the fortified enclaves or gated communities of the rich. It is precisely in Zona Sul, in areas like Morumbi (which lies on the same side of the Pinheiros river as Capão Redondo) that we see some of the most striking examples of contrast between extreme poverty and extreme wealth living side by side, pho- tographed and analysed by Caldeira. Racionais's strong emphasis on *periferia* corresponds, in the first place, to what Caldeira herself concedes remains after all the predominant spatial division of São Paulo, in spite of recent change. It corresponds, too, to the fact that Capão Redondo and its surrounding areas (as opposed to the Zona Sul more generally) are among the poorest parts of the city, remote *periferias* that can be considered to conform, still, to a more traditional division of space. And finally, in the poetic output of Racionais and other rappers – which has helped to define, as is argued here, an idea of neighbourhood subscribed to by many living in Capão – *periferia*, the Zona Sul, and Capão Redondo stand together in stark opposition to the richer areas of the city. As Imani Perry points out, rap is an aesthetic of oppositions (2004: 47), and São Paulo rap is no exception. But while some US rap defines opposition by territorial demarcation between gangs, its São Paulo counterpart concentrates exclusively on the contrast between the rich and middle classes, on the one hand, and the poor and *favelados*, on the other, inscribing that division onto clearly marked territories that correspond most of the time to actual territorial divisions. Yet this division typically acquires purely political and poetic meanings. Guasco, for instance, calls attention to the fact that the female hip-hop group Lady Rap, who lived in a run-down apartment-building in the very centre of the city, referred to their place as *periferia*. The use of Zona Sul as a synonym of *periferia* is another example.

LYRICS WITH REASON

The mapping of Capão as megalopolitan *bairro* in this sense is most thoroughly done in the words of the rappers themselves, in the rich lexical repertoire they invest in their songs, which enriches received definitions of 'lyrics'. Beginning with this concern with place, these texts define and evince understandings in the neighbourhood of basic factors such as class boundaries, drugs, marketed status, local heroes, violence, street status, race, and *bairro* solidarity. This is exemplarily the case with Racionais, where the actual means and methods of production of their work, scratching, for instance, or complex interplay between song, speech, the language it belongs to, and adduced 'noise', give unexpected resonance to the arguments proposed in the lyrics and to their cultural reference. This ranges, most often with razor-sharp irony, through giants of Greek cosmogony, Biblical chapter and verse, Hollywood, and *négritude*.

Working in from a *periferia* that is ontologically ever bound to be periferia, the 'rationality' announced in their name is urged (to quote some of their song titles) on the *bairro* jury, so as to broaden the limited mind of the 'Negro Limitado', and to invent a formula that brings peace like magic ('Fórmula mágica da paz'). Its reasonable language ransoms the 'active voice' of the grammarian ('Voz ativa'), barbarizes orthography to indict delinquency in 'Vida Loka' (reducible to 'V.L.', slang for a criminal). It also relies on enough English to rhyme 'playboy' with 'Hey boy', rehabilitate the term 'hood', and oscillate between the horror of Vietnam and the blues and browns of the lead singers.

Although we can see a gang culture operating in graffiti or tagging groups, São Paulo rap, on the contrary, moves frequently between neighbourhood specificity (like Capão Redondo) and a more general view of *periferia* that is ostensibly inclusive and implies solidarity. They repeatedly call attention to the similarities between different *periferias* – as the title of Racionais songs states 'Periferia é periferia (em qualquer lugar)' (*Periferia* is *periferia* (anywhere)).[5] The track 'Trutas e Quebradas' (Brothers and Hoods) reels off an extensive list of names of *periferias* in Zona Sul and other parts of São Paulo, and even other cities and states, followed by names of people and rap groups, messages to particular friends, etc. The long list is divided between the three MCs and is recited to no beat, just the background of an angelic female voice, humming a melody.

The richer areas of the city are rarely described. They are not called *centro*, nor are they usually referred to by their names. What characterizes them in the rappers' discourse is wealth markers: the existence of clubs and parks, of paved streets and expensive cars, restaurants, and mansions. Their most invoked inhabitant is referred to by the English term *playboy* (also found in the shorter form *boy*): the white boy from a privileged background. There are no good or innocent 'playboys' in rapper discourse here. In 'Hey Boy', for instance, the rappers spot a playboy driving a car in their neighbourhood. They noisily stop him and tell him to go away:

Hey, boy, what are you doing here?
My neighbourhood is not your place and you are going to get hurt
You don't know where you are
You fell into a snake's nest and I think you've got something to explain
It won't be easy to get out, life here is hard.

That is, the speaker opens by telling the *boy* to clear off because he will not be able to endure life in the neighbourhood, not having been trained, like the locals, to survive there. Later, the accusations become much more direct and unforgiving:

You've always had everything and did nothing for nobody
If things are bad, it's your fault too
Your parents turn their back on the world that surrounds them
They keep the bigger the better
and for us, nothing is left?
You spend fortunes dressing in brand names
While in the gutters children, men of the future,
eat almost nothing, die of hunger,
they're cold and afraid, it's no secret no more
and they take drugs
Feel the contrast and just agree with me.
Don't say no more because there's no point.

The song is played in a lively, almost jolly rhythm, much at odds with the sombre lyrics. Towards the middle of the performance, the singers start to distort the word 'boy' in the refrain, repeating it several times and making it sound like the Portuguese *boi* (ox). In other words, they mock and make fun of the *boy*, turning him into cattle or herd animal, someone who simply follows along having no mind of his own, like the white woman known as *vaca*, literally 'cow'. The mockery turns unambiguously into anger and insult at the very end, when the *boy* is called *otário* (stupid) and told aggressively to get out. What we hear in 'Hey Boy', in a fast, dance-like rhythm, is a clear demarcation of territory.

Even more directly accusatory is the brilliantly chilling 'Otus 500', which in invoking the violent giant that brought 'civilization' to pagan Greek makes its own contribution to the quincentenary celebrations of the 'discovery' of Brazil. As a comment on the 500 years inaugurated by the arrival of the Europeans in America, it emphasizes, in implicit solidarity with Brazil's Indians, that nothing has changed since then in power terms, and ends up feeding the paranoia and fear of the middle classes:

Yeah, mister, your titanic has sunk
yesterday's prey
has become today's predator

he got tired of being naive, humble and quiet
he got hooded became a bandit won't leave things as they were
Had a fit and went to the street to get his own
Check it out he's not opening your frigobar
in your living room
watching a DVD
with your wife as a hostage waiting for you
He wants to leave the plywood and live in a mansion
with a swimming pool good enough for a boss.

'Fim de semana no Parque' (Weekend at the Park) plays with the term *parque* (park), which, like its English equivalent, may refer to green or recreational areas in richer parts of the city, but which also, with perverse irony, has become a frequent toponym in some of the poorest areas in São Paulo (Parque Ipê, Parque Santo Antonio, etc.). The song moves between the two social uses of *parque*. One hour from the singer's neighbourhood he spots:

a brand-new car
all loaded, a guy's driving it
his children on the side
they are going to the park
All excited, electronic toys.

and also a club:

look at that club, how cool!
look at that court, that field, look
Look how many people
there is an ice cream parlor, cinema, heated swimming pool
Look how many 'boys'! How many chicks! Drown that bitch [vaca] in
the pool
There is kart racing, we can see
Just like what I saw yesterday on TV
Look at that club how nice!
Look at the Black young boy seeing everything from the outside.

The importance of the gaze in this song (and in Racionais's work in general) could hardly be overstated. Observing everything from a bus, the poetic 'I' is no neutral *flâneur*. In a consumerist society marked by such deep social contrasts, the gaze of the poor who look at the rich necessarily becomes (at least in Racionais songs) an expression of desire, resentment, or revolt.

Meanwhile, back in the neighbourhood:

thousands of houses piled up
muddy streets, this is the *favela*

my hood awaits me . . .
Here I see no sports clubs for the children.
No motivation! Sparse investment on leisure
the community centre is a joke
But then if you want to destroy this is the place
There's always alcohol and cocaine.

The weekend in this kind of 'park' is marked by violence:

I am tired of all this shit
of all this insanity
alcoholism, revenge,
betrayal and dirty tricks!
Anxious mother, problematic son
Destroyed families, tragic weekends
This is what the system wants, the kids just have to learn
Weekend in Parque Ipê.

Once again the song stands in clear contrast to its lively rhythm, and sounds especially ironic thanks to the sampling of Jorge Benjor (a composer who started with the 1960s generation, who fuses samba, funk and other pop rhythms) as he sings in his rather soft voice 'Vamos passear no parque, deixa o menino brincar' (Let's walk in the park? Let the child play).

In the highly concentrated poetic world of Racionais MCs, the richer and poorer parts of the city cannot and will not interact, except in violent terms. Like 'Hey Boy', other songs, too, make it clear that each party (the *playboys* and the brothers) should stay in its own place. In 'Da ponte pra cá' (On This Side of the Bridge), the border is clearly established by the bridge that crosses the Pinheiros River, although, as we saw, the river itself does not work so neatly as a social marker: Morumbi, one of the richest *bairros* in São Paulo, is on the same side of the bridge as Capão, while Brooklin, sung by another rapper named Sabotage and claimed as part of Zona Sul, is on the opposite side.[8] In 'Da ponte pra cá', as in 'Hey, Boy', strangers are told to be aware. In the refrain, sung with distorted effects that make it sound like a chorus of children, we hear: 'No use just wanting, you've gotta have some exchange, life is different on this side of the bridge'. Typical of Racionais songs, this one describes the streets of south side at night, from the point of view of Mano Brown:

Full moon illuminates the streets of Capão
above us just humble god, isn't it? Isn't it?
good health, women, and lots of music
white wine and a good lawyer for everybody
cough cough, hey, it's fucking cold
Tuesdays are bad for a stroll where am I going to go?

this has never changed and never will
the smell of bonfires dominates the air
same sky, same postal code on the south side of the map
always listening to rap to make the brothers glad
on the streets of the South they call me Brown.

Brown and his musical partners walk as *flâneurs* through the neighbourhood street, but rather than the urban modernity of a city centre, what they describe is the violent reality of *periferia*. Again, in the same song and now under the influence of marijuana ('smoke philosophy in the breeze'), he describes each *favelado* as a 'universe in crisis', and more desperately, he claims to understand how the desire to possess the same commodities owned by the rich may drive the *favelados* to crime and death:

who doesn't want to arrive in a black Honda with leather seats . . .
I never had a bicycle or video-game
now I want the world like Citizen Kane
this side of the bridge above all is a school
I want to get 10 as a mark
9.5 is no good
half a point below, brother, you're dead
half right does not exist, it's a common saying around here.

As we can see, the fact that Racionais songs establish a clear division within the city does not mean that they view the neighbourhood as an isolated place, turned in on itself. On the contrary. Not only is the neighbourhood part of *periferia*: it connects with other oppressed places of the world, as well, via rap and music in general. It is also portrayed frequently as the result of global economic processes. In other words, their construction of an idea of neighbourhood is at once deeply local and dependent on international phenomena, a megalopolis phenomenon.

In social and economic terms it should come as no surprise that groups such as Racionais MCs define their local space with the help of global media: the international music market is not the only aspect of global economy that young people in the *periferias* are involved with. As in many other cities of Latin America, the lack of job opportunities and a dysfunctional education system are driving young men to work in the drug trade, an underground and illegal industry that uses local labour but depends entirely on well-organized, international networks that make ample use of modern technologies and communication systems. In most places drug traffic is also connected to the illegal traffic of weapons: in the *favelas* of most Brazilian cities, gang and drug-related deaths often happen with the use of extremely modern and sophisticated weapons that can reach those localities thanks, again, only to well-organized international networks. As Racionais put it: 'Who sells drugs to whom? Hein? / It arrives here by plane or through the port or pier / I don't

know any poor folks who own airports and stuff / I get sad to know and
see / that those who die everyday are people like you and me' ('Periferia é
periferia'). Moreover, many of these youths die in pursuit of consumer goods
that are iconic representations of globalized capitalism: brand-name tennis
shoes and clothes, expensive watches, and cars. To quote Racionais again,
in 'Negro Drama': 'Hey Master / I know well who you are / alone you're
afraid / You said what was good / and the favela heard it / whisky, Red Bull,
Nike tennis shoes, machine gun.' Thus, in social terms, these young men belong
both to extremely disadvantaged communities, which often lack basic services
such as underground sewage, and to modern, international networks that
are, albeit illegally, a definite part of international capitalism. This is the main
theme of Racionais's songs and of many other hip-hop groups in São Paulo.
In fact, as happens with hip-hop in other parts of the world, their songs are
not just socially aware, they are didactic, and its socially transformative aim
is in itself a theme for many songs. 'Negro limitado', for instance, declares:

> Do it for your own sake, not for me
> Keep a distance from easy money
> too much drink, policemen and that shit
> At last, in an efficient way
> Racionais declare war against those who want to see Blacks in shit
> And the brothers who listen to us will understand
> that information is more important than a loaded Beretta
> Expensive brand clothes are worth nothing
> when compared to a well articulated mind.

Again and again the songs tell stories about criminals who had become local
heroes in crime and died, or who had got rich in the drug trade and died young.
The message to the young could not be clearer: in 'Tô ouvindo alguém me chamar'
(I can hear somebody calling me), one of their most famous pieces, a young
man tells his story while he is dying of a bullet wound. He talks about his
friendship with someone called Guina: 'strong partnership was the two of us /
high high high we got / we sniffed like mad (fuck) no sparing / all very cool.'
Guina ended up in prison and ordering the speaker's murder, on a false claim
of treason. Sung in a low voice and monotonous rhythm that is modified only
by narrative sound effects (shots, shouts during the robberies, vocal threats
at the moment of death, beeps from the pulse-counter, etc.), the narrative moves
brilliantly back and forth in time between different memories, made ever more
dramatic by the proximity of death. The speaker is unaware he is dying, while
the listener knows. Recalling one of his gun-point robberies, he declares:

> I only wanted to be cool, nothing else
> Show my brother in the hood
> a car and a fancy chick
> some money would solve my problem

what am I doing here?
my tennis shoe stained with blood, that guy on the floor
a child crying and me with a gun in my hand
It was a horror picture and I was the author.

The violence described in all the songs is the fruit of the current, very globalized context of drug and guns traffic, and of a post-industrial jobless society, no less a reality, of course, in other parts of the world. As Mike Davis puts it, pinpointing the year of especial interest here:

> Since 1980, however, economic informality has returned with a vengeance, and the equation of urban and occupational marginality has become irrefutable and overwhelming: informal workers, according to the United Nations, constitute about two-fifths of the economically active population in the developing world. In Latin America, adds the Inter-American Development Bank, the informal economy currently employs 57 percent of the workforce and supplies four out of five new 'jobs'.
>
> (2006: 176)

This situation is aggravated in Brazil by long-standing social disparity and by the lack of options given to the descendants of its very large slave population. As Racionais portrays it: 'there was a sinister *coup d'état*: the unemployment machine is producing criminals' ('Crime vai e vem' / Crime comes and goes). There are, nevertheless, other sources for the violence in their songs. Although Racionais often declare their hatred for globalized media, their songs, as Guasco observes, are clearly influenced by Hollywood films, crime programmes on TV, and video games: 'The scenes of explicit violence can be found in the *periferia* but also on television and the video game machines one sees in every corner bar, usually surrounded by a group of boys' (2000: 186).

If the richer areas, rarely described, are the scene of bitches, playboys, waste and ostentation, the neighbourhood home is portrayed in overwhelmingly negative terms:

> Smell of sewage, if it rains it is a nightmare
> a bit of hell is where I am
> Even IBGE [Institute of Statistics]
> passed by never to return
> They totted up the shacks and disappeared.
>
> ('Homem na Estrada' / Man on the Road)

Alcoholism is more prevalent than schools:

> Many open bars. Many empty schools . . .
> Mothers crying. Brothers killing each other. Until when? . . .
> Here, brother, it is each one on his own.
>
> ('Periferia é periferia')

Drugs are everywhere, and there is an ever present sensation of fear:

> Everyone is afraid of going out at night.
> Lately there are stoned guys on the streets
> completely high
> they rile you on their trips.
> ('Periferia é periferia')

Violence is experienced at a young age:

> I see this all the time, since I was a kid
> Already at fifteen
> treason in the party, lots of gunshots!
> Death, don't even mention it.
> ('Rapaz comum' / Regular Guy)

When Brown thinks of a house with a lawn and a swing hanging from a tree, a mocking chorus brings him back to reality, comparing his dream to a video game:

> Wake up brother
> This is Capão Redondo
> Not pokemon
> South side is the reverse, it is concentrated stress
> A wounded heart by square metre.
> ('V.L')

Violence is so prevalent that it becomes part of the landscape:

> Death here is a natural thing
> Common to see.
> Fuck! I don't want to have to find it normal
> To see a brother covered in newspaper!
> It is bad! Suicide routine!
> ('Rapaz Comum')

The street is where it all happens, and where the rappers can meet, talk, fight, and observe local reality. As Guasco (who has spent time on them) observes, the unpaved streets of *favelas* in Capão and other parts of *periferia* serve as a main social space (2000: 155; see also observation by Fenianos 2002: 75). The lack of leisure facilities and the small houses, which usually do not allow the luxury of privacy, force young people on to the street: they hang out at corners, or sit on the ground or on car boots, talking, listening to music, and smoking (or using other drugs). In Ferréz's *Capão Pecado*, for instance, the most important meeting point is the lamp post, followed by the improvised soccer field in an empty lot. Bars (mostly non-regulated corner bars) are of course another important focus for social life, but along with ball rooms, churches, and schools, they hardly appear in Racionais songs. Their songs do criticize the ubiquity of bars, and 'Você viu aquele mano na porta do bar' (Have you noticed that brother at the bar door?) is framed as a dialogue between two people who observe someone on the threshold between bar and street, and comment on his decline since he started dealing drugs. But even this song takes the street, not the bar, as the point from which the brother's life can be observed and commented upon. The street works therefore as a space of solidarity between brothers, but also as a 'school' where the rappers can teach each other and other brothers what life examples they should follow, or avoid. In addition, the street is the main stage for violence, for confrontation between classes (as in 'Hey Boy'), and repression, where the police and the vigilantes (adversaries more active than the *playboy*) create a permanent state of panic, as in the song 'Pânico na Zona Sul' (Panic in the South Side). Finally, the street is the space where rap and hip-hop are born, in their terms the *cultura de rua*, that is, street culture (Andrade 1996: 158; Guasco 2000: 154). Hence, the street functions for the rappers as a space of resistance, which they can create and change through their art.

Besides the street and the bar, the other public spaces most often invoked by these poets are the cemetery and the hospital. The first, specifically Cemitério São Luis at the edge of Capão, is described as being always full of bodies of companions and friends who have died and are now corpses, or who lie there wounded. Song after song tells of young men who thought they could escape violence but didn't. In the 1997 record, *Sobrevivendo no Inferno*, Mano Brown repeatedly calls himself a winner for having managed to beat the statistics: to have survived till the age of 27. One of the songs on that record, 'Tô ouvindo alguém me chamar', is the first person story of a moribund young man (see above). Another ('Rapaz Comum') is told from the point of view of a young man, already dead, who observes his own funeral. A third is the diary of someone in jail, and so on.

The same negativity and pathos pervade Férrez's novel, *Capão Pecado*. One by one all the young friends who would meet every day by a lamp post in Capão Redondo fulfill their tragic destiny. Capachão becomes a cop and ends up seen doing a search in the *favela*, touching up girls who had previously been his friends. Testa, who, like Will and Dida, had become addicted to crack, is killed by Burgos, the most sinister character in the novel, and so is Burgos's own brother, who is killed for being HIV positive. Burgos ends up being killed by a member of his own gang, and all of his partners end up, as well, in the morgue. Rael, the protagonist who to a certain extent represents Férrez himself (like Férrez, he is a bit heavy, wears glasses, likes to read and write) leaves the *favela* in order to work in a local factory. But tragically his wife Paula runs away with the factory owner, taking his son with them. Desperate, Rael becomes an alcoholic, kills his former wife's lover and is murdered in prison by Burgos's cousin. Matcheros, Rael's best friend, ends the book as the owner of a little workshop whose business seems to be the theft of motorcycles. No one escapes this bitter view of the neighbourhood: the future, for all of those young men, does not exist, or is most sombre.

Many of the narratives in the website capao.com also reveal the same closeness to death, and the same general bleak view of the place, as we shall see. According to Guasco, who did ethnographic research among Capão Redondo rappers:

> The negative view of *periferia* permeates almost all the lyrics of rap, it is present on the images that are shown on the covers of the cds, and was confirmed by my fieldwork, through its recurrence in several dialogues and interviews. Trying to exemplify it would be redundant, since rap is now well known by the public opinion and this is exactly the part of their discourse that is perceived outside rap's creative universe.
>
> (2000: 102)[9]

What is most surprising is not so much that the rappers should express such a negative view of their own neighbourhood, but rather that this negative view should be combined with an absolute and uncontested pride in being from there. Earlier in the same study, Guasco tells an anecdote that well illustrates what such pride may denote. Being a teacher in Capão Redondo, he invited several of his students to a party at his house. At a certain point during the party he put on a Racionais MCs record:

> While the introduction was being played, before Mano Brown's voice could be heard, the young men started to hop and scream frenetically, repeating several times in unison as loud as they could: 'The guy is from our place, the guy is from our place' . . . It was not only the fact that someone very famous came from a place so relegated to poverty and all of its consequences. Along with pride in the illustrious neighbour, there was also the pride in the fact that their place was being sung in his music.
>
> (ibid.: 100)

Mano Brown himself explains in his collaboration in Ferréz's *Capão Pecado* his sense of pride:

> Capão Redondo is poverty, injustice, dirt streets, open-air sewage, bare-footed children, crowded police stations, morgue vans coming up and down, tension and smell of marijuana all the time.
>
> São Paulo is not the wonderful city and Capão Redondo, on the south side of the map, much less so.
>
> Here crime stories are not romantic and have no heroes.
>
> So what! I love this fucking place.

In the big world I am nobody, but in Capão I have my place guaranteed, got it, brother?

(Ferréz 2000: 24)

And Cascão, another rapper from Capão who collaborated in Ferréz's novel, proclaims:

That's why we are high-on-drugs Capão, rapper Capão, outlawed Capão, illiterate Capão, everything Capão, even, get it, Sin Capão [the novel's title]!
But you know what, we are for love, and for love we'll die embracing Capão, because we're all Number 1 from the South.

(ibid.: 56)

The expression 'Number 1 from the South' is a reference to the movement created by Ferréz in 1999, which, forging the number '1' into a collective self, attempted to foster pride in coming from this poor, marginalized area of São Paulo. The expression appears on the tee-shirts (worn by young men) seen in the photographs of Capão Redondo which complement the text of *Capão Pecado*. It was this same pride in being from Capão that prompted the creation of the website capao.com.

Although São Paulo is generally described in the rap songs and in Ferréz's novel as an ugly city, the attractions of the richer areas are not ignored. That is clear in the song 'Fim de Semana no Parque', which, as we saw, compares the parks and clubs in the richer areas to the lack of leisure alternatives in the South side. The character Zeca, in *Capão Pecado*, makes the same type of comparison:

Zeca got the beer and went on drinking, but all of a sudden he remembered an article he had read that morning. It was about São Paulo being one of the most exciting cities in the world, one of the few that was awake 24 hours a day. In the article they talked about nightclubs, restaurants with all kinds of food that one could find at night. Zeca compared all that the playboys could enjoy and what he had here in front of him. He decided to stop thinking about it, walked a few metres and went to eat a kebab at Mrs. Filó's stand.

(ibid.: 43)

Yet the rappers, as well as Rael and his friends (from *Capão Pecado*) do not like to go to other parts of the city, and if they dream of sharing the life of the wealthier it is, as in Racionais songs, by way of a sarcastic and violent threat. When, in *Capão Pecado*, Rael's mother asks him to go to Liberdade (a central neighbourhood), her son cannot hide his displeasure: 'He was disgusted by those upturned faces, they seem to think they were better than other people' (ibid.: 35). And only on the way back, when he starts to come back to Capão Redondo, can he once again feel comfortable: 'He took the first bus, got off at the Capelinha station and took the Jardim Comercial bus. As the bus went its way, he felt better, he felt at home' (ibid.: 35). Supporting that view, the young Renato, a contributor to the website capao.com explains why he did not want to move with his father to Butantã, a middle-class neighbourhood: 'that place has only got playboys, it didn't look like me' (capao.com).

The feelings expressed here diametrically oppose those of Carolina Maria de Jesus when she wrote her diary of life in an East-side *favela* in the 1950s: 'When I am in the city I have the impression that I'm in a living room with crystal chandeliers, rugs of velvet, and satin cushions. And when I'm in the *favela* I have the impression that I'm a useless object, destined to be forever in the garbage dump' (1992: 29). Her return home, after being in other parts of the city, is always bitter. The term she uses here, and most often, to describe its relation to the rest of the city is 'junk room' (also the title of her book in Portuguese), that is, the room in a house where cleaning supplies and useless or broken objects are usually stored. Another term is 'ulcers': 'These views enchant the eyes of the visitors to São Paulo who never know that the most famous city in South America is ill with ulcers – the *favelas'* (ibid.: 77).

Carolina wanted to leave the *favela*, her great dream was to buy the masonry house which would become the title of her next book, and the theme of her tragic life. The richer parts of the city offered her a promise of liberation: 'When I go into the city I have the impression I'm in paradise. I think it just wonderful to see all the women and children so well dressed. So different from the *favela*. The different-coloured houses with their vases of flowers' (ibid.: 77). The images she uses to describe the relation between the city and the *favela* are coherent with her dream: the junk room is, after all, part of the house, and the wounds are part of the body. Both, junk room and wounds, could be changed, cleaned, healed. This is how she sees, as well, racial relations: racism, which she often denounces, is in her view a product of lack of education. Human beings are all equal, and one day they themselves will understand that.

In Capão Redondo, where most houses are already masonry-made, there is basically no way out, and if there is, it is not synonymous with abandoning the neighbourhood. For the rappers, for Ferréz, and for a good number of writers in the capao.com website, the *favela* and their neighbourhood are hell, Vietnam, war, terms which reflect perhaps the most obvious change in the everyday life of *favelas* in the four decades that separate the two books: the exponential rise in the rate of violence. Yet it is the home to be just as violently proud of. Racionais's song 'Homem na Estrada', tells the story of someone who dreamt of getting rich and leaving ('I want my son not even to remember this place'), but ended up being involved in crime and dying. Many of their songs tell the same story, also a common theme in Hollywood mythology, of men who got involved in crime and want to perform just one last time in order to get rich and leave, but end up dead. The moral message in these songs is obvious: involvement in crime is always dangerous and there is no possibility of getting out alive. But in the case of Racionais songs, they also point to another theme, which is the impossibility of leaving the neighbourhood. In 'Fórmula Mágica da Paz' (Magic Formula for Peace), Brown confesses to having sometimes felt the wish to leave:

> This shit is a minefield
> How many times have I thought of throwing myself out of here
> But then my hood is everything I have.
> My life is here and I don't need to leave.

This attachment to the place is not linked, as we have been seeing, to any intrinsically positive characteristics of the neighbourhood. It is true that in 'Fim de Semana no Parque', Brown allows himself one of the few positive descriptions of Capão, when he says:

> There is shouting at the front [of the bus], 'We are arriving!'
> I like this: more human warmth
> In the *periferia* there is joy as elsewhere
> It's almost noon, there is general excitement!
> It is there where my brothers, my friends live
> And the majority here looks like me
> I am bam bam bam I give the orders
> People have been dancing *samba* since ten in the morning
> Pay attention to the beat, attention to the tone
> 'What is it like, Mano Brown?'
> that's it: the number one in low income in the city
> South side community is dignity.

This song prompted Guasco to conclude that the rapper's supposedly negative view of Capão, and of *periferia*, is actually not so. *Periferia*, according to him, may be described negatively but it is also the place where there is more solidarity, and this would justify the rappers' pride in being from there. Strong as this song indeed is, it is, however, a lonely exception. In general, Racionais's songs do not show *periferia* as being a place where there is more solidarity. On the contrary: in many songs they complain about treason and betrayals, as in the same 'Fim de semana no Parque': 'I am tired of all this shit / Alcoholism, revenge, betrayals, tricks'. In 'Vida Loka (parte 1)', the speaker claims not to trust the streets:

> If someone is coming towards me, who is it? who will it be?
> bro, pass me my toy that perforates sweat shirts
> Because the guys who see me with my brothers around
> They try to see it, they want to know, they won't get it from me
> 'cause trust is an ungrateful woman
> she kisses you and hugs you, robs you and kills you.

In 'A Vida é um desafio' (Life is a Challenge), the poet complains that in the *periferia*

> the attitudes of the evil ones influence the good minority
> dying too easily killing too easily
> . . .
> it is unaccountable, unacceptable, implacable, inevitable
> to see the poor subjecting themselves to crumbs, favors
> sneaking in through the night for fear and horror
> what's the deal, the trick, the scene
> we pray we run away the same problems remain
> women and money are always involved
> vanity ambition: ammunition to create enemies.

In 'Crime vai, crime vem' (Crime Comes and Goes), the poet complains: 'this place is a den of thieves, a serpent's nest'. 'Jesus chorou' (Jesus Wept) is about gossiping and betrayal in the neighbourhood. In 'Expresso da Meia noite', the speakers openly declare: 'today no one trusts no one / me not even my shadow'. In 'Periferia é periferia', the rappers complain about local drug-addicts who rob hard-working neighbours at gun point, and steal clothes from people's lines. In 'Fórmula mágica da paz', Brown comments on what he calls the local code of not denouncing criminals, and says, bitterly:

> Unto each place its law, I know
> in the far South of the South side it is all wrong
> our law is flawed, violent and suicidal
> They say, they tell me that no one will tell
> is the first paragraph in law of the *favela*
> cool . . . what's scary is to discover that
> it is all for nothing and only the poor die.
> We are always killing each other, brother, why?

In fact, as we can see in these multiple quotes, the neighbourhood in general is marked by a persistent lack of solidarity among other negative character-istics, which again prompts the question: why be so proud of being from there?

The reasons are not linked to any intrinsic beauty of the place, nor to the fact that the people there are better (because according to the songs they are not), but that they 'look like me', that is, they share the same ethnic and social origin. In the words of Mano Brown in the song 'Negro Drama', even if you leave the *periferia* it does not leave you: 'Money may remove a man from poverty / But it cannot remove the *favela* from inside him'. This notion of a place where the rappers and characters feel at home, where they belong, is related, of course to the family (almost exclusively to the idea of the mother) but it is defined above all by the group of friends, the 'brothers' whose sense of community comes from the fact they all 'look like me', that is, they do not belong, almost ontologically to the wealthier parts of the city. This separation, or apartheid, is defined in social terms (the poverty, which, in the words of the *Capão Pecado*'s protagonist Rael, goes from parents to children), but also racially. The US rap groups that influenced Racionais with the idea of a socially committed rap, also inspired them with a conception of race relations that is at the centre of their idea of neighbourhood. Brown says as much in an interview. When asked what had been his main inspiration for his rap, he replied:

> Public Enemy. When I read Malcolm X's book, it was the same time. I almost went mad [because of] the stuff about race, brother. Race, black, white, some stuff he said that happened there [in the US] and you see it happening here in exactly the same way. You think: shit, the man is telling the truth, he is not telling lies.
>
> (Pimentel interview)

In this sense, most *paulista* rappers see themselves as part of an African diaspora that has more connections with former slaves and the oppressed in other parts of the world than with the Brazilian elites – confirming, once again, Gilroy's hypothesis. This conception of race relations has been derogated as an import, as having no connections with Brazilian reality, where, according to the same critics, race relations are of a different order.[10] The rappers, meanwhile, are very conscious of their own motives in adopting such view of race relations. At the end of 'Tempos difíceis' (Hard Times), for instance, amid howls of laughter a voice proclaims 'Brazil is country with tropical climate where races mix and there is no racial prejudice'. Their importing of race ideas from groups such as Public Enemy or 2Pac, or from US Black intellectuals like Martin Luther King and Malcolm X, is not done as mere repetition: they are certainly aware they are bringing this discourse into a context that has tended to reject it, and they do so in order to provoke that rejection. They want to raise awareness about how prevalent racism is in Brazilian society and, along with that, pride in being Black.[11] Many of their songs are addressed to Blacks who fail to recognize their own Blackness, such as 'Juri Racional', which explicitly foregrounds the concept they chose as a name for: 'You will be judged for playing in the opposite team / for dressing like a playboy / going to bourgeois dance clubs [danceterias] / *négritude* is what it's about in the end'. If the idea of race is indeed inspired by US popular culture, it is critically placed in the Brazilian context. The song 'Voz Ativa' (Active Voice) discusses how carnival, once a Black Brazilian festivity, has been taken over by commercial interests, that relegate the Blacks themselves to marginal positions:

> White models on the floats
> Where are the Black women?
> They dance on the ground, at an inferior level

> No originality, every year more commercial
> Carnival used to be the people's party
> It used to, but some Blacks sold out again
> Whites above, Blacks below, all this is normal, natural.

'Capítulo 4, Versículo 3' (Chapter 4, Verse 3) starts with Mano Brown reading, without any musical background, the following statistics:

> 60 per cent of young men in the *periferia* without any criminal history have been abused by the police.
> 3 in 4 people killed by the police are Black.
> In Brazilian universities only 2 per cent of the students are Black.
> Every 4 hours a young Black man dies a violent death in São Paulo.

It is only when that fact is understood that one can find, as it were, a way out. In *Capão Pecado*, the protagonist Rael comes to an epiphany when he discovers that he could not escape his race:

> Rael tried to think about God, but he asked himself what heaven was like
> . . . would there be a *periferia* there? What about God? Would he belong to the master's mansion or would he live in the slave quarters? He understood it was all wrong, that the heaven they show is elitist, the almighty and cruel God they hide has killed millions; it is in the Bible, it is there, he thought, but they present Jesus as a blond guy. What the fuck is that, what kind of pattern is it? He then got to the most obvious conclusion: this is hell, where we pay and are paying, this is the hell of somewhere else and since the *quilombos* we've been paying, nothing has changed.
> (Ferréz 2000: 73)

Though there are no other options for any of the characters in Ferréz's novel,[12] alternatives do appear in the rappers' vignettes that initiate each chapter, and in their songs. The alternatives are not in so-called honest work, though that is acknowledged to be preferable to drugs. They are to be found in studying and reading, not necessarily in school nor as a form of getting a job, but as a way of learning about one's place, one's history and one's race; and in creativity, that is, rap, dancing, painting, writing and playing sports. This is how Ferréz's character Narigaz explains to his friend Matcheros how they can compete with the playboys: 'Know what I mean? The playboys have more options, but in my opinion we have to beat them with our creativity, you with me? We have to destroy the sons-of-bitches with the best thing we've got: our talent' (ibid.: 118). In Racionais's songs, rap is presented again and again as the best alternative, sometimes as the only alternative to drugs and criminality. It is actually through creativity that a sense of community is established. The neighbourhood is portrayed by the rappers as all they have and the only possibility of building something of their own. The songs constantly sing the streets, describing them, as we saw, as horrible and violent, but also and constantly as their own place, the place they want to change through their songs. Poets in the songs appear frequently listening to rap, playing rap, and self-referentiality, in an autobiographical sense, is a constant characteristic of rap, not only in São Paulo. When desperate after almost losing his life because of a betrayal, Mano Brown declares:

> for my brothers I have my presence to offer
> maybe it is a bit confused but it is real and intense
> my best Marvin Gaye
> Saturday on Marginal [a road that follows Pinheiros river, and is therefore called "Marginal Pinheiros", linking North to South side]
> whatever will be will be
> and we'll go to the end
> let's connect, wherever it is necessary
> in paradise or in the Pastor's judgement day
> connect me and my brothers is what I ask
> *favela* Fundão [a specific favela in Capão Redondo]
> immortal immortal in my lines.

('Vida Loka, parte 1')

In 'Fórmula Mágica da Paz', Brown lists different parts of the neighbourhood, speaking as a now successful singer who tries to give back to the community he still belongs to:

> I try to guess what you most need
> build your house
> buy some clothes
> a lawyer to get your brother out
> on the visiting day you tell me
> and I will send cigarettes to the crazy ones in jail.

As urban rhythm, rap centres musically on the idea of producing neighbourhood, by bringing in the surrounding 'noises'. Racionais songs do exactly that through the incorporation of street sirens, discussions, screeching tyres, conversations, children playing, telephones, the police, and the 'noises' that is, the music that is sampled, changed and distorted by the scratching techniques. These are, too, part of neighbourhood noise, as indeed are the radio, which appears frequently in the Racionais songs, TV, and other electronic media, and dialogue. The MCs talk among themselves, telling each other stories, discussing various themes, boasting, teasing each other, sometimes agreeing, sometimes not. Many of the songs recreate dialogues with people in the streets, in bars, on the telephone, enhancing, through the multiplicity of voices, the sense of place. In 'Vida Loka, parte 1', for instance, the poetry is interrupted to incorporate a dialogue with passing friends, inviting them to play football. In 'Capítulo 4, Versículo 3', Brown cruises the neighbourhood listening to rap in his brown 'opala' (an old Chevrolet), calling some friends by their names, challenging them to play pool or dominoes. In the process he includes some people he does know: 'two brothers waved at me / wearing satin jacket and jeans'. Brown then is told not to talk to them:

> Hey Brown, cut it out, don't even try
> no use talking to these guys
> last night I saw them near the paved road
> inhaling death blowing life away
> look at the guys just flour [cocaine] skin and bone
> at the bottom of the pile / the evidence in their pockets.

Brown replies: 'look here, nobody is better than nobody / look here, they are our brothers too'. In 'Jesus chorou', Brown gets a call from someone who wants to tell him that he heard somebody speaking badly about him, saying that he talked all the time about the neighbourhood and *periferia* but actually: 'what *periferia*? he only thinks of himself / full of dough and you guys eating shit'. The voice also says that all this was said in public, but nobody in the group had responded at all. Though there is obvious malice in the friend's voice and the intonation shifts, implying a lie, and that for his part Brown does not believe what he is hearing. He shows a superior attitude: 'those who have a mouth speak just to get a name / to call attention from women and men', but in the privacy of his bed, he cries, like Jesus when he was betrayed by Judas. In other words, the format of rap, its inclusive style of dialogue and street noise is fundamental to the process of producing place, because it accommodates voices in disagreement, different points of view. Perry sees this inclusiveness of other voices and opposite views as one proof of the democratic tendencies of rap (2004: 7). In one of Racionais's best songs, with a mock biblical title, 'Capítulo 4, Versículo 3' that translates as 'Record 4, Song 3', Brown speaks simultaneously as himself and as a criminal (though a criminal that uses words, not guns), emphasizing thereby the contradictions that are crucial to the idea of neighbourhood that they are building.

It is this mode of creating a neighbourhood through art that endows them with a sense of belonging. What is positive about the neighbourhood in the rappers' songs (and also in the vignettes of them in Ferréz's novel) is not that the neighbourhood is a place with more warmth and solidarity, but that they are creating a space of solidarity among themselves, in the community they are making through their own work. This space of solidarity is not free of contradiction, or even betrayal: it is a work in progress, a process that is still developing.

WORK IN PROGRESS, AND ON LINE

Such emphasis on creativity and invention is strongly confirmed by the website capao.com. Many fields on the website are dedicated to the local heroes, rappers, writers, footballers who, thanks to their fame, are lights on the horizon. Indeed, as already mentioned, most texts consist of histories of hip-hop and home-grown groups, especially Racionais: interviews with Mano Brown and other successful rappers, commentaries on hip-hop songs, reviews (positive) of books by Ferréz and interviews with the author. Texts sent recently to the website constitute an archive rich in information on matters of current concern in Capão, and alert us to taxonomy and practical questions of how best to lay out the screen, and to the range of functions the site may have socially in the *bairro*. A sampling made in last winter in São Paulo (May–June 2006) brought up submissions which, counter-posing rap lyrics (and the prose of Ferréz), cover such topics as music and the loyalties of those who make it, confessional reports by women and men that reflect on how the sexes behave towards each other, foundational rhetoric of how the site came into being, the life stories of those who contribute to it, and more.

Leonardo Lopes's description of the musical group Banda Tecora provides a convenient point of entry. After classifying their musical genre as 1960s sound mixed in with 'ghetto music', Lopes goes on to make a strong spatial claim: 'They have their feet firmly placed in the *periferia*, the southern end of São Paulo, Capão Redondo, but have a global view of the poverty and deprivation that most of the Earth's population is subject to' (capao.com). That is, Lopes hears Banda Tecora as both rooted in Capão Redondo and belonging to a global economy that causes 'poverty and deprivation'. Local identity, in other words, is created through a global discourse, not in opposition to it.

In language that is as vernacular as it is grandiose, Alexandre M. O. Valentim reviews the names of Capão's famous offspring, in the article 'Periferia: berço de valores' ('*Periferia*: Cradle of Values'). He notes that those who have made a name for themselves in Capão Redondo have had to 'overcome prejudices, resentment, lack of support, bad economic conditions and all kinds of obstacles', and wonders what more could have been done had there been 'minimum support for and investment in these worthy people?' (capao.com). The names he lists are Mano Brown, André Luiz de Souza (film-maker, now resident in France), Ferréz, Natarja Turetta (actress), Cafú (footballer), Lucila (city representative), and Father Jaime Crowe (an Irish priest who fights for human rights). Most of these names come from the arts and sports, swelling the emphasis on creativity as the best (and sometimes only) job option for young people in the *periferia*. The website's emphasis on hip-hop, other sorts of music, and writing, reinforces that idea, as indeed do many of the social initiatives that it describes and promotes: soccer and hip-hop workshops, and nothing less than a 'fábrica de criatividade' (literally 'factory of creativity').[13]

On the one hand this emphasis on 'creativity' reflects the social/economic predicament of current capitalism, where jobs are scarce and badly paid, leaving young people in poor neighbourhoods with little hope of a future. It reflects, as well, the specific situation of places like Capão Redondo, where the majority of the young people are black and will have to face racism when trying to get accepted in the job market. In the Racionais MCs' song 'Negro Drama', Mano Brown comments on this lack of options, being ironic about his own success: 'Crime, football, music / I didn't manage to escape that either / Count me in' (*Nada como um dia depois do outro dia*).

Yet, it is precisely this creativity that allows these young men to forge their own idea of place. Elias Lubaque, an avid contributor to the website, quotes fellow citizen Roberto Teixeira Barbosa, who says:

> living in a *bairro* like Capão Redondo may be dangerous but it is also gratifying because the *bairro* is making history and gaining space and a voice in society after almost 100 years of existence. It is not much, but we have to fight to achieve more. Artists and other people who are respected in the 'system' came from here, and this no doubt is a sign that things are going to change.
>
> (capao.com)

Capão is 'making history', that is, it is becoming a recognized place, a well-known neighbourhood, because of the rappers and writers, because of the 'creativity' of its citizens. It is not only that Capão inhabitants can now feel proud because their place is home to famous and successful people. More specifically, the famous and successful people from Capão, particularly the rappers and writers, sing about Capão, write about Capão, live in Capão, and are socially involved in Capão (this of course is just the mechanism detected early on in Tepito by Monsiváis when he wrote about the famous boxer from the barrio: see Chapter 4). In other words, they are actively engaged in producing their own idea of Capão. They do this not by offering conventionally positive views of the neighbourhood. On the contrary, their strategy is to turn their own weakness into strength: by talking about crime, violence, and drugs, they manage to attract attention from other sectors of Brazilian society at the same time as making themselves understood by the local youth. They repeat the much propagated view of Capão as a violent, dirty and ugly place, yet still reaffirm their pride in being from there.

This process of creating their own geography actually has consequences that go beyond the local. Rap's aggressive lyrics, its use of neighbourhood slang and supposedly 'incorrect' Portuguese (see below), its 'noisy' interference in urban environments unmistakably launch an attack on the city and Brazilian society as a whole. Ferréz's characters often use the term 'revolution', the revolution that will come when the brothers recognize 'we are the majority, we just have to use our heads, studying, getting informed and waiting for the big turn. When it arrives, we're going to say: yes, *favela*, take over from one extreme to the other' (2000: 161). Brown and Ice Blue sarcastically comment on the consumption of their music by the playboys. Addressing the rich and powerful with the slavery term 'Plantation Master', they say (as any good Jesuit might in the war for the soul):

problems with school, I have a thousand,
a thousand deals,
incredible, but your son imitates me,
amongst you lot,
he is the clever one,
he swings and speaks slang,
no, not slang, dialect.
this one is not yours anymore,
look at him
gone
I came in through your radio
took him,
you didn't even notice.

Not only are their songs (and Ferréz's literature) overtly didactic and political. Their authors, too, have become political leaders in their daily life, contributing with what they do (as well as with their art) to consolidating Capão
Redondo as a neighbourhood. They are involved in social projects in Capão
Redondo. Mano Brown campaigned in favour of gun control in the 2005
vote, and was, prior to Lula's election, a fervent supporter of the Workers
Party (PT), bringing party members to Capão Redondo and trying to get
local people to vote for them. Ferréz refused to sell the rights of his *Capão
Pecado* to a film producer because, he claimed, 'lately they are opening stores
there and this generates jobs. But the investment was going to stop if they
made a violent film about Capão' (Giannetti). He also recently criticized
Rio rappers MV Bill and Celso Athaide for having allowed Globo TV to
produce a TV documentary on their book *Falcão: Meninos do Tráfico*
(Falcão: Drug Traffic boys. Ferréz: 'Antropo(hip-hop)logia'). In other words, as
they have become public figures frequently present in the mainstream press,
they have been forced to take the consequences of their own artistic work,
and strive to forge their own idea of space in new political arenas.

The cultural production of these groups from Capão Redondo neatly
confirms Edward Soja's assertion:

> If traditional equality politics mobilizes its radical subjectivity most
> expansively around taking collective control over the 'making of history',
> then the new cultural politics of difference, identity, and representation,
> without lessening the power of its historically inspired strategies, adds
> a new source of mobilized consciousness rooted in the more immediate
> collective struggle to take greater control over the making of geography
> – the social production of human spatiality. This involvement in producing
> and in already produced spaces and places is what all those who are
> oppressed, subordinated, and exploited share, and it is this shared
> consciousness and practice of an explicitly spatial politics that can pro-
> vide an additional bonding force for combining those separate channels

of resistance and struggle that for so long have fragmented modernist equality politics.

(2000: 281)

There is no doubt that their creation of a geography is, in all senses, a very conscious movement of resistance, and an attack on the profound social and racial inequalities of Brazilian society. At the same time, it is important to remember García Canclini's criticism of those he calls naïve neo-Gramscians in *Hybrid Cultures*:

> The difficulties become more acute . . . when their models are used as superparadigms and generate popular strategies to which they attempt to subordinate the totality of the facts: all that is not hegemonic is subaltern, or the inverse. The descriptions then omit ambiguous processes of interpenetration and mixing in which the symbolic movements of different classes engender other processes that cannot be ordered under the classifications of hegemonic and subaltern, modern and traditional.
>
> (1995: 99)

First, there is the question that haunts most successful rappers and pop musicians around the world, which is their actual and highly integrated participation in the consumer society they so harshly criticize. This has been a recurrent theme in rap for many years,[14] and Racionais, following many of their US counterparts, do not avoid the issue, touching on their own contradictions at the same time that they use the image of the successful black man as another form of resistance:

> So, at the time of the plywood shacks at Pedreira [a *favela*]
> where were you?
> What did you give for me?
> What did you do for me?
> Now you've got an eye on what I earn?
> Now you've got an eye on what car I drive
> It took a long time but I want more
> I want to have your soul.
>
> ('Negro Drama')

Also, in spite of the fact that these young men all came from *favelas* and are writing, in that sense, from within, it is also true that the very fact that they are writing and have gained access to publishing and media puts them in a different social position from the rest of the neighbourhood. Again, this issue has been raised many times by rappers and critics in the US. It touches on the question of 'authenticity', that is, of which rappers have the right to represent (or continue to represent) the neighbourhood once they are successful.[15] Brown is very clear in this respect: at the end of 'Negro

Drama', he assumes a loud declamatory tone to declare that: 'I didn't read it, I didn't watch it: / I live the negro drama, I am the negro drama, / I am the child of the negro drama'. In a less histrionic mode, the clearly didactic aims of their art puts them in the position of 'educators', and consciousness-raisers for the rest of the community – the position that Brown also discusses, as we saw, in 'Jesus chorou'. As George Lipsitz claims in the case of Chicano rock, these rappers and writers have become, in a Gramscian sense, 'organic intellectuals' whose aim is to raise consciousness in the rest of their group (quoted by Forman 2004: 215).

More complicated still is the issue of gender, a notoriously thorny subject when the subject is rap. Many articles and books have been written attacking or attempting to justify rappers' frankly misogynist and homophobic songs. Those of Racionais, unfortunately, are no exception to that tradition. Their song 'Mulheres Vulgares' (Vulgar Women) rehearses discourse well worn in gangsta rap, about women who want men only for their money. Women, especially white middle-class ones, are billed as bitches (*vacas*) and whores. True to the label Cosa Nostra, the song 'V.L. parte II' clearly spells out the mafioso ruling that 'my friend's woman is the same as a man' (i.e. sexually off-limits), blaming women who seduce and betray for behaving like men, and going to the extreme of saying that a man who sleeps with a friend's woman deserves 'a bullet in his face'. Women in their songs never speak in slang and their Portuguese is much more 'correct'. They belong, in that sense, to another world, as it were, and returning the compliments paid to them have no respect for the rappers. It is important to note that critics are always willing to be implacable towards rappers with regard to misogyny, forgetting, as it were, that our society (and in this case Brazilian society) is deeply (if less overtly) misogynist as whole. As Mark Anthony Neal observes:

> As sexism and misogyny are largely extensions of normative patriarchal privilege, their reproduction in the music of male hip-hop artists speaks more powerfully to the extent that these young men (particularly young black men) are invested in that privilege than it does to any evidence that they are solely responsible for their reproduction.
>
> (2004: 247)

It is also important to take Tricia Rose's very fair point that the attacks on women often reveal an insecurity (typical of younger males) about women's sexual powers (1994: 171–3). At the same time it is difficult to understand why the degree of sophistication that Racionais display in the analysis of most aspects of Brazilian society, and the very sophistication of their songs, should not continue to operate or be as subtle when they are dealing with the position of young women in their own neighbourhoods. The problem is twofold: first, it is obviously not true that women as a whole are interested only in men's material possessions. Second, if some women are indeed interested in these possessions, why can they not receive in these rapper songs the same treatment as, say, young men who fall into crime? In other words, like young men in *periferia*, young women anxious to prosper or just survive have very few options besides the pursuit of richer partners, or prostitution. If it is true that such aggressive treatment of women betrays sexual insecurity, it is also true that it reveals other forms of social insecurity, related, for instance, to the role of consumption in their urban environment, as well as – we have to admit – adherence to conservative patriarchal values.

Given the fact that more than 90 per cent of the victims of violent death in the Brazilian *periferias* are young men, and that young men are also the main perpetrators of those murders, one could say that drug violence in Brazilian (and Latin American) cities is gendered. Of course women are also victims of violence in these neighbourhoods, but rarely do they suffer violent death.[16] The general ethics behind this violence is also masculine, appealing to ideas of honour, manly courage, and bravery before death. Racionais texts like 'Mulheres Vulgares' and *Capão Pecado* blame women for the fact that men are tempted into crime and die so young. Their message to the brothers is clear:

> Don't fall into this trap
> Be smart with the world and pay attention to everything and nothing
> Women want or prefer only what favours them
> Money and possessions, they forget you if you don't have it
> We are *Racionais* [rational], different, if not the same
> Vulgar women, only for one night and nothing more.

In a society where they are displayed everywhere (seldom fully clothed) in billboard advertisements, magazines, and television, it is not surprising that women are seen as a consumer product, a product that, like Rolex watches and expensive cars, may drive men to commit crime in order to obtain it. But women are also consumers, and one of the commodities that they consume is men: what the Racionais song and Ferréz's story complain about is precisely the fact that women can dispose of men when they find them useless, that is, when they no longer deliver the goods. In other words, what makes men desirable as commodities, according to these texts, is the fact that they enable women to obtain more and more commodities. Men are, in that sense, the ultimate commodity. What seems to haunt these authors would then be the fear of being turned into a mere consumer item by women. No wonder the Racionais recommend the brothers not to allow themselves to be consumed and, instead, be the consumers: Vulgar women only for a night and nothing more. For the rappers, as for Ferréz, the antidote to being consumed by women appeals to a somewhat vague idea of 'true love', to soap-opera morality.[17] If society has been dominated by consumerism most thoroughly in the megalopolis, not least its *bairros*, the women who live there are expected to become the guardians of disinterested relationships and 'true love'. The fact that they refuse to play that role and become consumers, like men, is precisely what is unacceptable for those men.

We should also add that these young men live in neighbourhoods dominated by the drug traffic, whose strict rules and absolute demand for loyalty
they end up emulating. Hence their appeal for listeners attracted, even if
subconsciously, to the morality of the underground drug world.[18] Their songs
create a masculine utopia based on solidarity between friends (fear of treason,
demands for loyalty) that in the end comes to resemble, in spite of clear
ideological differences, the ethics and codes of the underground drug trade.

In Ferréz's second novel, *Manual Prático do Ódio* (A Hands-on Guide to
Hatred), we witness an attempt to change this view on gender issues. For
one of the members of the group that is planning a great robbery is a woman,
Aninha. Yet quite unlike the men in his stories, she is a highly idealized
character: she dresses in man's clothes and is at the same time very sexy and
feminine; she is experienced and courageous though somehow still naïve. She
is also the only member of the gang to escape death (she returns to her native
North-east). Sex in the novel recalls that described in Ferréz's first text: a
violent relationship where the power is always on the side of the man.[19]

The website capao.com reinforces many of the gender stereotypes found
in Racionais's songs and in Ferréz's novel. But the multiplicity of voices that
characterizes the new and ever-increasingly frequented medium of the internet allows for the inclusion of a more complex view of the relationship between
the sexes. Simone, for instance, tells a story of violence that appears to be,
at first, a mere repetition of the ones quoted before: Patrícia and Rafa meet
and fall in love but Rafa betrays Patrícia with her best friend and the two
grow apart. Patrícia becomes pregnant by another man while Rafa becomes
a criminal and goes to prison. Patrícia visits him frequently and hopes to be
with him again but when he is released he returns to the world of crime. Patrícia,
in the meantime, dies, hit by a stray bullet, and Rafa laments not having seen
her again. The difference here is that the story is told from the point of view
of the woman, 'a young mother of just twenty-four whose only desire was to
live a great romance', but who ends up being a victim of violence (capao.com).

Marco Garcia in 'A invenção' ('The Invention') lists the numerous ways
in which women are superior to men, while Karen Dias overtly attacks
machismo in the form of men who get drunk and treat women badly.
Ananda, a young hip-hopper, reveals a determination that goes far beyond
consumerism. Having been frustrated with the rest of her group because, after
much preparatory work, the others lost their motivation to send their music
to a contest in Florianopolis, she reaffirms, in beautifully stylized *paulista*
slang, her own desire to go on making music. And in the story by Fábio
Calvisio Rodrigues it is the woman who saves the young man from crime.
In this way, capão.com reworks the highly masculinized community elaborated by Racionais and by Ferréz by opening up a dialogue with feminine
and feminist voices.

Needless to say, this does not mean that the website is inherently more
liberating, as a medium, than traditional media like music and literature;
capão.com voices many opinions about gender-related issues, and some are

palpably more hide-bound than others. Yet, as a neighbourhood website for all, capao.com opens up possibilities of expression that simply did not exist before.

In its current format, the homepage of capao.com has at its centre an explanation and history of the site itself. On the screen to the left we can see a list of collaborators and, at the top, a list of topics: articles, events, our art, humour, dialect, 'o comédia e o truta' ('the clown and the buddy', a service for posting public messages to friend and foe), and 'stories from the ghetto'. There is an overlap between authors and topics: a link to an author name posts a complete list of her/his contributions to the site, which can also be reached via links to topics (such as stories from the ghetto, or humour). Not all collaborators claim to be from Capão, and some give no personal details. Among those who do describe themselves, we can see certain variations in their social and educational backgrounds: some collaborators have gone or are going to college, while others have had little formal education. For the first time in the history of the West, people who would never before have had access to publishing have gained access through the internet, and the texts they write are reaching impressive numbers of readers. As Mitra puts it:

> The Internet has transformed popular culture by providing a virtual forum in which different communities and groups can produce a 'presence' that might have been denied to them in the real world. This presence can be obtained in cyberspace by appropriate use of one's voice to articulate the specific narratives and discourses about ones group or subculture.
>
> (2004: 492)

To be sure, none of the collaborations posted on the website to the moment exhibit the sharpness, analytical and poetic skills (not to mention musical skills) of Racionais songs; nor, indeed, the powerful repetition of a deadly everyday life as described in *Capão Pecado*. As an anthology or digest, however, they demonstrate that *narrative* plays an important role in the creation – to quote Soja again – of a local geography, in this case that of the megalopolitan *bairro*.

In the section of the webpage called 'Stories from the ghetto', most of the narratives talk about events that are very similar to those described in the rap songs and in Ferréz's novels: young men who get involved in crime and drug trafficking and are killed; or who nearly get involved and escape; or who know somebody who got involved and got killed. Tales in this vein are many, and include Eder Paulo Kogus's short text 'Será que tenho anjo da guarda?' ('Do I Have a Guardian Angel?'):

> Saturday night at some friends' house celebrating two bro's birthday some-thing came up and I took a chick to my house three hours later I went back to the party and when I got there where is everybody just blood on the floor bullet marks on the walls God knows what happened yeah

it seems bullshit but if I were there I had been shot and some people still complain about their lives.

(capão.com)

Fábio Calvisio Rodrigues's story, 'A vida não é conto de fada' ('Life is Not a Fairy Tale'), relates how he became involved in crime after his parents separated but was saved by a girlfriend.

Some authors talk not about themselves but people they claim to know: Ricardo Barbosa Silva, for instance, tells a brief story about someone he knew as a child. 'D' was a 'nice kid but a bit passive so he was bullied by everybody'. When the author was already working as a *motoboy* (messenger or delivery boy on a motorbike) with a helper and a girlfriend, 'D' started 'to steal, go off the rails, go up the wrong alley, robbing cars, buses, transporting kilos [of drugs] from place to place, driving powerful bikes, GM Corsas, VW Golfs, and me always poor, telling him: brother don't do that, man, life is short, let's enjoy it the right way, no way' (capao.com). Not surprisingly, 'D' died as a result of choosing the 'wrong life'.

Jonas de Oliveira tells a very similar story about a childhood friend with whom he used to 'goof around in buses' and root together for Corinthians (one of São Paulo's four main soccer teams). While Jonas kept his job and started a family, Sabão, his friend, started to hang out in very bad places and ended up murdered. 'Unfortunately this happens every day on this side of Pinheiros river', he adds (capão.com).

Other contributors prefer invented narratives to real life stories. Another Simone (Ferreira), for instance, in 'Vidas sem sentido' ('Lives without Meaning') creates a symmetrical story about two young men, José and Fernando, who live respectively in the *favela* and in an expensive condominium next to it. They become friends. Fernando gets attracted by the idea of drugs and crime which the *favela*, for him, represents; while José (whose deceased father had run a drugs ring) wants to study and 'be somebody'. In a Manichean social inversion, José becomes a successful lawyer and Fernando, after spending some time in prison, ends up killed in a bank robbery. It is interesting to note that while Jonas de Oliveira (in his 'real life' story) subscribes to the rappers' division of space that follows the course of the Pinheiros river, Simone Ferreira in her invented story creates a geography that much more resembles the one described by Teresa Caldeira in *City of Walls*.

Marcos Lopes in 'Por trás de uma grande mulher' (Behind a Great Woman) narrates a story in the first person about a young and humble construction worker who could not attract women. After helping someone at the roadside whose car had broken down, he was invited to a party and introduced to Juliana, with whom he instantly fell in love. At first he lied to her about his profession but when he could no longer afford her expensive tastes, he confessed his humble condition, and she left him. Heartbroken, he fell into a life of crime and became very rich and successful. His father would accept no money from him and his mother harped on how his old friends, poor and humble, were managing to study and get jobs. In what he describes as his 'last robbery' his friends were killed by the police. He escaped, but is now narrating the story in prison, to a lawyer.

The message arising from both the 'real life' narratives and the invented stories repeats itself and coincides, too, with those coming from the hip-hop songs and Ferréz's novel: every young man in Capão and the *periferia* lives too close to drug trafficking and crime and is going to be tempted to join in, at some point in his life. But deliquency never pays, always ends in early death and should be avoided at all costs. Although some of these stories claim to be real and others fictitious, all of them are in one way or another narrated from the point of view of survivors: writing, after all, can be done only by those who have survived. Read as a collection, these stories provoke the same feeling one has when reading Ferréz's *Capão Pecado*. In the novel a young man becomes an alcoholic after being abandoned by his wife, kills her lover and ends up being killed himself in prison. This melodramatic central plot is, as it were, superfluous, since the strength of the novel lies in the beautifully monotonous repetition of the young men's attempts to stay alive. The main character in the web narratives is, as in the novel, the attrition of everyday life in the São Paulo *periferia*. Through the repetition of similar incidents and personal tragedies, the web stories, along with the rap songs and Ferréz's novels, create a sense of community that is at once virtual and geographically bound. It is as if the contributors were saying to each other: 'I can understand your story because I live in the same place as you, and because I live there, I can also tell a story that is similar to yours.' Together, they form a kind of 'hypernarrative' which tells the life and death of a young man from the south side of São Paulo. The sense of place created by this 'hypernarrative' stems from the conditions of production and consumption in present-day capitalism: on the one hand, massive unemployment, especially for unqualified young men; on the other, massive pressure for consumption through globalized media. It stems, as well, from narrative structures found in different media: Hollywood films, radio and TV soap operas, romantic literature, Wild West comics, video games, etc. Rather than celebrate a proletarian hero, this 'hypernarrative' tells the tale of the unemployed young man of the megalopolis who sooner or later will be tempted to live a life of crime.

Here we find, once again, an important difference between the web narratives and the novels and songs to which they are so closely connected. The rappers and Ferréz focus on the stories of those who did not manage to escape crime, on anti-heroes who have fallen victims to an extremely unjust social system. Most of the website postings, on the other hand, shift the focus to those who did escape. They are in that sense epic stories, in which the decisive and heroic act is seen as choosing the 'good way' (studying, working, and in some cases converting to a religion) and resisting the temptation of falling into drug traffic and crime. In other words, in spite of its clear emphasis on creativity and rap, the website also makes it clear that one does not have to be a rapper to be a hero in Capão. At the same time that it works as a celebratory affirmation of the neighbourhood created by the rappers, and by Ferréz, the website's openness to a multiplicity of voices – a heteroglossia in Bakhtinian terms – allows it further to function as a re-writing of the artists' Capão into a more nuanced geography (one glimpsed it may be surmised by the improbable *urbenauta* Fenianos when he was welcomed into the *mutirão* (see Chapter 3)).

One aspect of both the website and Racionais's songs (and not so much in Ferréz's novel) that asks for further future studies from a specialist is the effect they are having, and will continue to have, on oral and written language. Racionais's songs brought *paulista* youth slang to millions of people (mostly young) of all social classes throughout the country – an unprecedented feat. Of course, the oral dialect from the city of São Paulo had been given plenty of artistic expression before, in the works of Galvão, Machado, and Juó Bananare (noted above), and São Paulo's great modernist, Mário de Andrade. The line carries on through Carolina Maria de Jesus and João Antonio, and the *sambas* of Adoniran Barbosa in the 1950s (which acquired a certain comic folk reputation) and the *paulista* punk movement of the late 1970s and 1980s (which had, however, little national repercussion). It is not only that Racionais songs use terms from street and crime life, or idioms born from rap and break dancing. The grammar, too, is the characteristic grammar of oral *paulista* language, which eliminates the plural agreement between nouns, pronouns and adjectives. In 'correct' Portuguese, the phrase 'the yellow houses' would translate as 'as casas amarelas', the 's' at the end of each word being an indication of plural. In typical *paulista* dialect, the plural is indicated only by the article: 'as casa amarela'. The social prejudice against such language is immense. Omitting the plural is traditionally considered a marker of ignorance and illiteracy, which translates as indicator of poverty. Racionais brought this language aggressively into their songs, and made the most of it, creating and forcing rhymes that are part of the street language but that would not be possible in normative Portuguese.

The case of the website collaborations is even more striking. Since the postings on capao.com. are not edited or corrected, they give us a fresh feel of *paulista* slang and syntax. Surprisingly, perhaps, this vernacular varies little between people who claim or appear to have minimal formal education and those who boast a college degree. Exceptions are rare, and involve one or two contributors who write professionally for local newspapers and blogs. The overwhelming majority of the texts reveal the characteristic suppression of plural markers and no grammatical agreement between subject and verb. Language purists would probably argue that the generally lower educational level (especially in private *periferia* colleges), or the influence of popular culture as in the music of Racionais is destroying Portuguese grammar. It could more encouragingly be seen as (added) proof of the strength and resilience of speech, of people hearing each other intimately even in the megalopolitan wasteland. Due linguistic analysis of these contributions would require specialized study that goes well beyond our purpose here. But the fact remains that for the first time these speech traits are gaining a significant written register. It is of course too early to predict how this alphabetic record might impinge on *paulista* language norms and literature, but it is hard to imagine that either will remain unaffected. Already it is redefining the difference between the 'two languages of São Paulo' – the spoken and the written – famously observed by the character Macunaíma, eponymous hero of Mário de Andrade's ground-breaking novel of 1928.

As neighbourhoods, Capão and Tepito alike have in the past few decades been defining themselves in terms that carry the story of the very large city into another phase. This is exemplarily so in all that concerns the invention and defence of their own space, in facing difficulties unprecedented in the western experience of community. Through discourse that is common to them, they create an imaginary of the *barrio/bairro* less readily reducible to current tenets of the social science, fundamental as these remain, or coincident with party programmes, beyond basic rejection of the right. With differing emphases in matters of land rights, race and gender, these megalopolitan places echo each other across the continent when, polite or not, they speak from within, re-police the polity, find global in local, re-centre western policy, affirm pride in rationality or deep historical memory, and wish urgently to survive.

6 Writing on the wall and other interventions
Epilogue in a small gallery

Life in the Megalopolis has no encyclopedic pretensions, and even if it had, it could never attempt to map, even nearly, the totality of cultural productions from the two Latin American megacities in the last quarter century. Fragmentary, as all representations of the megalopolis necessarily are, it concentrates on a few significant examples of 'megalopolitan experience'. As in the case of Ruffato's book, this format encourages us as well to think about what it does not include.

Originally, the book was meant to have a further chapter, dedicated to the subject of violence in Mexico City and São Paulo. It was only when I was well advanced in the writing process that I realized that violence did not need a special chapter of its own. As a theme, violence pervades all the chapters and all the works we have been looking at. The book's structure concentrates, on the one hand, attempts to account for the immensity of the two cities, and, on the other, the production of local geographies, and this has the effect, nonetheless, of pushing to one side works and phenomena that deserve mention, if only in passing. On the Brazilian side, Beto Brandt's excellent film *O Invasor* (2002), and Vinicius Mainardi's not so excellent *16060* (1995) both point to a phenomenon well worth considering with regard to themes of violence and class difference in São Paulo. Both resort, to varying degrees, to a somewhat farcical plot in order to bring the rich and the poor together, and the relationship between the classes is, in both cases, necessarily permeated by violence. It is as if realism or hyper-realism (salient in recent depictions of violence in Brazil, like the blockbuster film about Rio, *Cidade de Deus*)[1] was no longer enough, in the view of these authors, to represent colossal class difference. Or, to put it another way, it as if the gap between poor and rich were so immense that only farce could accommodate them in the same room.

As for the genre of the detective novel, it continues to thrive in the megalopolis in its own term. Works produced in the two cities integrate violence into what has been shown to be for us the enduring concept of the *flâneur*. According to Benjamin, the detective is 'preformed' in the *flâneur* (1999a: 442). Set in the megalopolis, novels by the Mexican Paco Taibo II and the *paulista* Tony Belloto tend to treat the detective, and crime in the city, in quite conventional ways. As a result, continuing to search earnestly for clues and to track down the criminal, these detectives seem somewhat out of place, even out of their depth. Especially in the work of Taibo II, urban crime is so pervasive that the hard-boiled individual detective, almost in spoof mode, is ruminating humorously on his role.

Dystopian science fiction, another predominantly urban genre, seems to find a home more readily in Mexico City than in São Paulo. *Leyenda de los soles* (1993), the novel by Homero Aridjis, takes place in a future Mexico City that needs water so badly that even the rain has dried to dust. Projection into the metropolis that awaits us, in this case the quincentenary year (1992) of Columbus (Cristóbal Colón) also characterizes *Cristóbal Nonato* (1987) by Carlos Fuentes. Both authors invoke Mexico's deep history in imagining their futuristic societies. Pre-Columbian gods and Aztec calendrical notions create a sense of time that is, in both cases, more complex than in standard science fiction. The very title of Aridijis' novel is that of a Nahuatl account of the 'suns' or 'world ages' inscribed on the *piedra de los soles*, excavated in Tenochtitlan's Templo Mayor. Once again, no equivalence for this temporal understanding of dystopia is to be found in São Paulo.

Perhaps one critique that might be made of the choice of works analysed in this book is that very few of them, if any, depict either megalopolis in a positive way (a possible exception is one or other of Bruno Zeni's poems). Violence, class disparity, racial and class prejudice, ecological disaster, and fear permeate all accounts of the two megacities studied here. As a result, the intense and exciting cultural life of both Mexico City and São Paulo (cited by many inhabitants as a reason for continuing to live there) is mostly omitted. Truly, cultural life is not nearly so common a theme in the literature, cinema, and even popular music in the two cities as are violence and class difference. But there are exceptions: Alberto Guzik, for instance, in his novel *Risco de Vida* (Risk of Life, 1995) and collection of short stories *O Que é Ser Rio e Correr* (What it Means to be a River and Flow, 2002) celebrates various expressions of high culture and night life in São Paulo: theatre, dance, literature, painting, classical music, and so on. Hugo García Michel's *Matar por Ángela* (Killing for Angela, 1997), whose title hardly eschews the idea of violence, conveys the reader from one part of Mexico City to another, to the fast rhythm of rock concerts. The characters of Juan Villoro's *Materia dispuesta* (Disposed Matter, 1996) are constantly going to parties, restaurants, openings of exhibitions. And Armando Ramírez's twin novels, *Sóstenes San Jasmeo* and *La casa de los ajolotes*, though both grim depictions of violence and corruption in Mexico City, nonetheless take the reader on a 'tour' of cabarets, restaurants, and bars in the capital in its 1950s heyday, and in the 1990s. The city features as a space of sexual freedom and gender liberation in the works of Guzik and Michel; Regina Rheda's short-story collection *Arca Sem Noé* (Ark without Noah, 1994), set in Edifício Copan (Oscar Niemeyer's 'S'-shaped landmark building in the centre of São Paulo); and in virtually all the stories by Maria Emília Kubrusly collected in *O Bar de Flora Paixão* (Flora Paixão's bar, 2000), to mention but a few.

Finally, how does cultural production modify or interfere in the space of the megalopolis? The previous two chapters showed us how literature, rap, and web literature are attempting to produce space in two of the poorest and most violent neighbourhoods in Mexico City and São Paulo. To close, I have set up a small gallery of visual works that endeavor, for their part, to impinge on the space of the two megacities.

RUINS

Although the urban area of São Paulo is still the largest industrial centre in the country, the signs of the de-industrialization that started in the mid to the late 1980s are everywhere. A familiar image in many areas of the city is the closed metal doors of abandoned industries and related commerce. Streets that once were bustling with movement of pedestrians now appear empty, occupied only by passing cars. This is the case, for instance, of Cambuci, a former Italian and working-class neighbourhood near the centre of the city. Or of the urban stretch of Via Anchieta – the road that links São Paulo to the coast – in Sacomã; or even of Brás, another Italian neighbourhood east of the centre, and so forth.

Visual artists have been particularly sensitive to this change in landscape. The project Arte/cidade Intervenções Urbanas, led by the artist Nelson Brissac Peixoto, has been attempting to draw attention to particular areas of the city that have suffered drastic economic change in the past decades. They do this via art installations that use the city itself as a place of exhibition. In their first event in 1994, for instance, they used the abandoned slaughterhouse of Vila Mariana. In 1997, the project named 'A cidade e suas histórias' (the city and its histories) chose three sites within an area of 5 kilometres: Estação da Luz (the central train station), the abandoned factory of Moinhos Santista, and another abandoned factory, Moinho Matarazzo. Visitors to the exhibition had to travel between the places by a train specially arranged for the exhibition (passenger trains are nowadays very limited in São Paulo). Some 34 artists took part in the collective exhibition in the three sites. In the description of the exhibition, Peixoto mentions the special place of industrial ruins in the contemporary city: 'The modern city is the stage of incessant transformations that reveal its precariousness. Ruins and construction sites are indistinguishable' (as in Claude Lévi-Strauss's sentence about Brazil, popularized in the song by Caetano Veloso, 'A Nova Ordem Mundial').

In her installation for the Moinho Matarazzo site, Laura Vinci created a mechanism that very slowly transferred one ton of white marble powder from the top floor of the ruined building to the bottom (Figure 6.1). A movable sculpture directed downwards (instead of up), the slowly un-forming and slowly forming piles of dust literally brought the movements of time (as in an hour-glass) and of active occupation of space (the moving sculptures) back to the abandoned factory. To use the Benjaminian expression, the installation created a phantasmagoria of work space through the machine that transferred

Figure 6.1 Laura Vinci, Untitled, Exhibition Arte/Cidade III (1997). Photograph by Laura Vinci.

dust from one place to another, and through the device that measured (as in factories) the passing of time. It served, as well, as an allegory of São Paulo's (ruined) industrial economy.

MORE RUINS

Artists like Laura Vinci and Nelson Brissac Peixoto intervene in the city through large projects that depend on grants, sponsorship, and the patronage of well-informed viewers. Graffiti artists, on the other hand, interfere in the space of the city with recourse to much less, and often actively harassed persecution by the police and many of the city's inhabitants. In São Paulo, and in Brazil more generally, there are two words for graffiti: *grafite* (or graffiti), which describes drawings (especially when they are considered artistic by the viewers); and *pixação*, used mostly to refer to tagging or to designs that are not seen as aesthetically pleasing. Needless to say, there is much room for negotiation between the two extremes. During the term of Luiza Erundina as PT mayor of São Paulo (1988–92), 'artistic' graffiti artists were invited to paint certain public spaces of the city, and to teach in public workshops. Roaleno Ribeiro Amancio Costa wrote a Master's thesis on the results of this initiative by the municipality, which is fascinating above all because it includes several interviews with *grafiteiros*. While some of the most successful ones defended their participation in the workshops by describing themselves as visual artists in a broader sense, hard-line *grafiteiros* accused the participants in the city's initiative of selling out, of being co-opted by the machine. Most importantly, these *grafiteiros* defended an idea of their art that was aesthetically tied to its illegality. Júlio Barreto, for instance, defines graffiti as 'the spirit of intervention. For me, graffiti is that, it is to interfere in what is forbidden' (Costa 1994: 88).

Now that the practice of exhibiting graffiti artists in galleries in São Paulo (and in many other cities worldwide) has become much more common, this discussion from the early 1990s can serve as an important reminder of some basic aesthetic principles of graffiti. Although not necessarily linked to mega-cities the size of São Paulo or Mexico, graffiti and tagging are both urban art forms *par excellence*: they use the space of the city as their canvas, and depend on not being seen while creating their work. Taggers actually divide the space of the city in territories that are disputed between gangs, and use writing as codes that serve both to identify groups and to restrict comprehension. They also use the space of the city as a challenge to be overcome, particularly in the case of those groups that enjoy writing on surfaces extremely difficult to reach. Taggers and graffiti artists often reject accusations of 'dirtying' the city by pointing to the visual pollution of advertisements as being equally or more offensive than their art; in other words, they see their interference in the city as a comment on advertisement, commodity culture, and pollution. As Barreto says: 'Who are the worst aggressors? The little graffiti artists who want to show off or the mega-companies? (ibid.: 91).

Perhaps even more significant in aesthetic terms, the creations of graffiti artists and taggers are dependent on the particularities of urban spaces: are the viewers going to be sitting on buses? Are they going to be driving fast or be stuck in traffic? Are they going to walk by? Will they look up? From what angle? A long series of images colourfully painted at the exit of the short tunnel between Paulista Avenue and Rebouças, for instance, can only be seen from a car: it is virtually impossible for a walker to reach that space. The creation of graffiti artists is therefore dependent on their deep knowledge of how people move and behave in particular urban spaces.

These considerations can help us look at the graffiti painted on an abandoned factory in the former Italian neighbourhood of Cambuci, photographed in 2003 by Fátima Roque (Figure 6.2). The whole building has been colourfully decorated with writing and human figures. In the central image, some

Figure 6.2 Fátima Roque, Untitled. Photographed in Cambuci, São Paulo, 2005.

people are trying to take off and fly in a balloon. Seen in detail, the balloon reproduces patterns of fabric in a mended patchwork, designed around an old air-conditioning unit. To its left, a closed metal door of the abandoned factory is completely covered with writing (illegible in the photograph). Many of the designs incorporate the decaying surface of the wall, but are also being destroyed by decay. In neighbourhoods like Cambuci, where industrial activity has ceased and industrial architecture is simply left abandoned, graffiti interventions like this serve to call attention to the abandoned space, so as not to let it disappear in the landscape. By incorporating the very decay of the surface and the old air-conditioning, for instance, and by transforming the metal layers of the door into lines to write on, *grafiteiros* 'recycle' the former industrial spaces. Moreover the theme of the old mended balloon is in itself related to the idea of old, discarded technology, even more so because it includes, as it were, the old air-conditioning.

THE DANCER: A GRAFFITI MURAL IN MEXICO CITY

Up to now, graffiti in Mexico City has been a phenomenon less studied than its counterpart in São Paulo. In both cities (and in most urban centres in the world), the drive to produce graffiti, the tagging gangs that colonize the city, the materials involved and the police harassment seem not to vary much.

Figure 6.3 The dancer graffiti. Photograph by Antonio Iannarone, 2003.

What differs is the fact that in Mexico City the art of writing on walls cannot be dissociated from local visual codes that preceded it. This is the case of the celebrated post-revolutionary murals which constantly invoke the word-image of pre-Cortesian *tlacuilolli* (here, no counterpart is to be found in São Paulo). An example is the mural graffiti of a dancer, photographed by Antonio Iannarone in 2003 (Figure 6.3).

The frame allows the viewer to see enough of the setting chosen for this mural painting to guess that it is a derelict lot. In itself, this marks a stark contrast with its main image, that of a refined and urbane dancer. The artists sign their names in a fashion comparable to the hieroglyphic signatures painted on classic Maya ceramics and to that degree associate themselves with the legendary refinement of that tradition. In his posture and attire, the dancer recalls figures seen in the Olmec-Maya murals that have been uncovered in Cacaxtla and disseminated during the last three decades. In this respect, the dancer can be thought to allude to notions of rebirth elaborated in the *Popol vuh* (the 'bible of America') and other Maya texts. At the same time and in line with the style of murals found as far afield as the San Francisco Mission, the image as a whole appeals more broadly to the visual language of *tlacuilolli*, and to Mexica representations of the dance through time.

The dancer's midriff is protectively girt with a snake body, coiled so as to resemble the ball-player's 'yoke', and which to either side projects to form gaping caiman maws, heavily armed with teeth. These appear to shield the dancer from mechanized threat of the modern city, the hard edge of unfeeling urban mass. On the Aztec Sunstone, such creatures protect gods, and embody complex understandings of time, being known in Nahuatl as the Protean time snake xiuhcoanahual.

The visual statement made on this mural, like those generated by the Revolution, has much to say in the modern megalopolitan context about concepts of time and survival that stem from millennial memory.

FRONT-PAGE NEWS FROM A TENOCHTITLAN ROOFTOP

The massive immigration that has densely populated the northern approaches of the megalopolis recalls for cartoonists the arrival from that direction of the Mexica or Aztecs who founded the city in 1325, brilliantly so in the Columbus quincentenary year of 1992. In that year, *El Ahuizotl* (the satirical supplement of *La Jornada* issued as a modest antidote to the triumphalism that officially characterized it) drew on accounts of the Aztec arrival preserved in the codices, written in *tlacuilolli*. Modern reality is inserted to great visual effect, as the immigrants turn up travelling now not along the footprint trail but on dilapidated buses, and so on. Captions reportedly supplied by Monsiváis uncharitably assign these arrivals to the tribe of the Nahuatl-sounding Imecas, a term in fact derived from the then recently introduced metropolitan monitor of air quality (Indice Metropolitano de la Calidad del Aire). One of *El Ahuizotl*'s cartoonists Rafael Barajas (El Fisgón) went on to compile *Cómo sobrevivir al Neoliberalismo sin dejar de ser mexicano* (How to survive neoliberalism without ceasing to be Mexican, 1996), drawing ever more ingeniously on *tlacuilolli* sources. Similar techniques and the same politics nourish the quincentenary work by Eduardo del Rio (RIUS), *500 años fregados pero cristianos* (1992, 500 rotten but Christian years). As if intuiting the imminent discrediting and downfall of the PRI (the 'Institutional Party of the Revolution' that had held governmental power for a lifetime), RIUS published five years previously *Quetzalcoatl no era del PRI* (1987, Quetzalcoatl was not a PRI person). RIUS takes his sources seriously, updating meditation on Mexico's most famous 'plumed serpent' Quetzalcoatl.

The news that grabbed the headlines in the inaugural number of *El Ahuizotl* (Figure 6.4) tells of this Aztec influx. Announcing this event in 1325, the front page focuses on an eagle holding a snake in its talons and beak,

LLEGARON DESDE AZTLAN LOS AZTECAS año 1325

SE CUMPLIO LA PROFECIA DE HUITZILOPOCHTLI

FUNDARON CHILANGOTITLAN

EN MEDIO DE UN LAGO

EL AGUILA DESAYUNANDOSE UNA CULEBRA FUE LA SEÑAL

COLOCARON LA PRIMERA PIEDRA DEL CENTRO HISTORICO

Otras Profecías:

bres blancos y barbones ... Quetzalcóatl, Coloncóatl y Cortescóatl, víboras
rán a celebrar emplumadas que vendrán un día desde el viejo
iinto aniversario mundo a enseñarles español

Figure 6.4 'El Ahuizotl'. *La jornada* (Mexico) March 4, 1992.

the sign taken by the Aztecs that they had arrived at the place where they
were to found their city. On Mexico's national flag and countless other official
logos, the foundational eagle has been consecrated as the central image. In
the codices the eagle alights on a cactus that sprouts from a stone (the *tlacuilolli*
toponym legible in Nahuatl as te-noch-titlan). Here, the eagle has landed all
right, but the cactus has been replaced by a TV aerial on what looks like an
ugly rooftop in the Centro Histórico. That is, the Tenochtitlan heartland which
local inhabitants claim they never left. The poverty they now live in and
the unwelcoming image they are made to project in the media lead to the
satirical thought in the accompanying text that the eagle would have done
better to have landed further south, say, in a rich *colonia* like Lomas.

LOMAS BELLES AT HOME

As Mike Davis (*City of Quartz*), Saskia Sassen and Teresa Caldeira have shown, one tendency of the contemporary megalopolis is violent separation, the ever-widening gulf between social classes. This may be epitomized not just by the severe degrees of poverty and marginalization shown in the works discussed above, but at the other extreme, by the 'fortified enclaves' in rich neighbourhoods, which consist of walled condominiums, apartment buildings guarded by security towers, private policing, 'armed response', and so on. These places, however, tend to figure less in cultural representations, presumably for the same reasons that make anthropological and sociological studies of the upper classes quite rare. The elites are not so easily convertible into 'subjects'; they have the means to project images of themselves when they want and to protect themselves against unwanted peeping whenever need be.

The result is that rich neighbourhoods as such hardly ever feature as a subject in the literature, cinema, and popular music of Mexico City and São Paulo. This is not to say that rich areas of these cities never *appear* in them, just that they hardly ever serve as subjects *per se* of cultural representations. This situation lends especial interest to a recent exception in the visual arts: the collection of photographs in the book *Ricas y Famosas* (Rich and Famous Women) by Daniela Rossell. It grew out of a solo exhibition 'housed in condemned mansion in the ritzy neighbourhood of Polanco' (Gallo 2004: 114). In it, young wealthy women, mostly from families of the top echelons of the PRI living in a very exclusive neighbourhoods such as Lomas, pose inside or in front of their houses, in postures and scenarios of their own choosing. The result is an egregious series of highly exoticized images of the rich, enough to feed the voyeuristic impulses of most common citizens. Though the book is not supposed to be about 'neighbourhoods', but about the young 'rich and famous' women, what strikes the reader is the constant revelation of what is inside some of the houses in those wealthy areas of the city. In other words, the photographs seem to be less interested in the women as such than in the space they inhabit.[2]

There is something in the photographs that reminds one of a zoo, not least the constant presence of animals, predominantly not local fauna. One woman poses sitting on a zebra skin, in a room also adorned by a stuffed spotted leopard (or jaguar?) (Figure 6.5). Another, inside an incredibly ornate room, dresses in attire fiercely at odds with the setting, ready for tennis complete with a racket, and extends incongruity by dangling her foot over a stuffed lion (Figure 6.6). Yet another appears wearing a white fur coat inside a large gallery adorned with a range of stuffed species that someone must have got past customs control, including a (polar?) bear (Figure 6.7). For one brief moment (the time it took for the shot to materialize) we are allowed a peep into the houses of the rich, and what we see in the case of these photos is not normal life, but very strange, highly exoticized venues.

Figure 6.5 Untitled. Daniela Rossell. *Ricas y Famosas*. Hatje Cantz, 2003.

Figure 6.6 Untitled. Daniela Rossell. *Ricas y Famosas*. Hatje Cantz, 2003.

Figure 6.7 Untitled. Daniela Rossell. *Ricas y Famosas*. Hatje Cantz, 2003.

It is difficult to know exactly what might have been the intentions of the photographer, herself a young woman from the wealthy areas of the city. She makes no comments, and lets the photographs speak for themselves. Nor can we be sure about what might have been the motives of the young models in posing as they do. As Gallo notices, 'They appear to be lacking something. They are not arrogant or haughty, snobbish or even wholly self-assured' (2004: 115). In any case, the bizarre exoticization of the rich in Rossell's photographs reinforces their extreme isolation, in these walled and heavily policed neighbourhoods (where common citizens and much less the poor cannot penetrate) from the rest of the city.

Notes

1 APPROACHING THE MONSTER

1. And you were a hard start / I put aside things I don't know / and coming from another, happy city dream / we learn quickly to call you reality / because you are inside-out of inside-out of inside-out of inside-out ('São Paulo', Caetano Veloso).

    For how long, on what omen-less islet, / shall we find peace for the waters / so bloody so dirty so remote, / so subterraneanly virtual now / of our poor lake and muddy / volcanic eye, god of the valley/ whom no one saw face to face and whose name / the ancestors never said? (Translated by Gordon Brotherston).

2. Including the current president, Lula, and his immediate predecessor, Fernando Henrique Cardoso. Lula was not originally from São Paulo but made his career as a union leader in the industrial suburbs of the city.

    Anthony Sutcliffe points out that most giant cities are capitals, New York being one of the few exceptions. São Paulo should be added as another exception.

3. These are the sites for the population data: Google Earth:
    http://bbs.keyhole.com/ubb/showthreaded.php?Number=598494;
    United Nations: http://www.un.org/esa/population/publications/WUP2005/2005WUP_agglomchart.pdf;
    City Populations: http://www.citypopulation.de/;
    City Mayors: http://www.citymayors.com/features/largest_cities.html
    http://www.citymayors.com/statistics/largest-cities-population-125.html

4. See also Bridge and Watson (2003: 104–12), and Amin (2000).

5. In spite of the large-scale de-industrialization, both cities remain major industrial centres in their respective countries (Damiani 2004: 26).

6. As in the title of the collection of essays edited by Philip Kasinitz: *Metropolis. Center and Symbol of our Times* (1995).

7. For guidance on Aztec Mexico, I have relied on Gordon Brotherston, especially his *Book of the Fourth World* (1992) and *Painted Books of Mexico* (1997).

8. For an excellent history of everyday life in revolutionary Mexico City, see Schell (2003).

9. For an excellent history of urban modernization and cultural modernity in São Paulo in the 1920s, see Sevcenko (1992).

10. In the 1990s, São Paulo copied from Mexico a system of pollution control that consists of certain cars not being allowed in central areas of the city during rush hours in one day of the week. The system is called 'rodízio' in São Paulo and 'no circula' in Mexico, and the choice of cars not allowed to circulate on a particular day of the week is based on the number plate.

11. See Brotherston's *Book of the Fourth World* and *Painted Books from Mexico*.

2 IN FRAGMENTS FOR THE MILLENNIUM

1. Unless otherwise stated, all translations of texts listed in the bibliography under titles in a language other than English are mine.
2. In this regard, see also the opening chapter of Donald (1999).
3. I owe this information to Marisa Lajolo, who obtained it directly from the author. She is currently preparing a study of Ruffato's work and kindly indicated that the changes in his life alluded to here had to do with a decision to free himself from a fixed place and hours of employment.
4. In his analysis of *Eram Muitos os Cavalos*, Flávio Aguiar proposes its constituent sections as the beads of a rosary, which concords with this notion of 'circular' time.
5. Directed by Pedro Pires and Zernesto Pessoa of the company Feijão, the play opened in 2003 and contrived the interconnection between the scenes and their locations with remarkable skill.
6. See the essay by Beatriz Resende (2001), for further discussions on this point.

7. The silent flow of the machines. Several buildings have little red lights at the top, like little points. To guide the airplane routes. From above the city is a field of yellow, white, and red lights. A sea of lights that twinkle. Other buildings have big white torches that light up when you pass by. One is walking on the sidewalk and when you pass by the sensor – usually installed in the doorman's tower – fiat lux. It is almost always a shock. A dive, too, because one doesn't see anything for a few moments.

8. See, for instance, Vice and Savlov.
9. The possibility of this seismic reading and its implication in geological time, raised in the French journal *Positif*, is noted by Smith (2003: 53).
10. In a paper presented in the Autumn of 2005 at Bolivar House, Stanford University, CA.
11. Paul Julian Smith notes that there was a house keeper in the original script, which reinforces my point even more.
12. According to papers in the film's archive, in an earlier version the *colonia* was explicitly said to be Condesa, a detail I owe to Smith.

3 FLÂNERIE

1. For connections between citizenship and contemporary cities, see Holston (1999).
2. Canal Once also airs Pacheco's *Conversando*, in which she interviews well-known names in the world of arts.
3. For more comments on Baudelaire and heroism, see Marshall Berman (1988:143–4).

4 BARRIO/BAIRRO

1. See Lira for a complete discussion of this point (January 17, 2007).
2. See Rogério Proença Leite, *Contra-usos da Cidade. Lugares e Espaço Público na Experiência Urbana Contemporânea* (2004).
3. According to some sources, *fayuca* started to come to Tepito in the bags of locals who went regularly to the US and sold the goods themselves. Over time, the activity became more organized and some *fayuqueros* started to specialize in the transport and selling of imported goods to the street vendors. Nowadays, the goods supposedly arrive directly from China (Rico 2006: n.p.).

4. We are the product of the Revolution.
5. Hernando de Soto is a Peruvian economist known for his enthusiastic celebration of the 'informal economy'.
6. Nowadays, after the PRI's loss of the presidency in 2000, many street vending associations are affiliated with the PAN and PRD.

5 CAPÃO REDONDO AND THE SPACE OF RAP

1. *Folha de São Paulo* 25 Jan. 2004.
2. The question of unauthorized and unregulated construction in São Paulo is discussed by Maricato (1996).
3. The past five years have seen an upsurge of this kind of literature in São Paulo. There are many prison diaries, for instance: Jocenir's *Diário de um Detento* (2001); André du Rap e Bruno Zeni's *Sobrevivente André du Rap* (2002), and Humberto Rodrigues's *Vidas do Carandiru* (2002). José Marques Sarmento's *Paraisópolis: caminhos de vida e morte* is a novel about the Paraisópolis *favela*, written by a local and electrician by profession. Ademiro Alves (Sacolinha), is an activist and writer from the East-side *periferia*, whose novel *Graduado em Marginalidade* (2005) was indicated for a literature prize in 2006 (he also writes for capao.com), and so on.
4. For a comprehensive history of São Paulo hip-hop see Andrade (1996) and Pardue (2004).
5. In recent years Rio de Janeiro has also become an important center of rap, above all through famous names such as MV BILL and Celso Thaíde.
6. See also Bennet (2000).
7. The title is actually a quote from Gog, a Brasília rapper.
8. Officially, Brooklin indeed belongs to Zona Sul, though the neighbourhood is in practice better known as an upper-middle-class area. It does, however, have poor enclaves in it, which justified Sabotage's appropriation of it as a Zona Sul *periferia*. Sabotage was a brilliant rapper, with a particularly fast way of singing, a lot of humour, and the tendency to incorporate samples from a great variety of rhythms. He was murdered at a bus stop in 2003, for motives still unclarified, not long after his participation in Beto Brant's film *O Invasor* (2002). For its part, based on the book of the same title by Marçal Aquino, this film opens powerfully announcing São Paulo to be a concrete monster in a process of contact construction.
9. For discussion of 'Crime talk' see Caldeira (2000). Pardue (2004) contrasts this negative view of *Periferia* with Evangelic 'Positive' rap.
10. A recent example of this type of accusation can be seen in an interview with the pop star Caetano Veloso in the monthly magazine *Cult*. In it, Caetano reports a discussion he had previously had with Rio rapper MVBill:

> [MVBill] was talking about a debate he had with [film director] Arnaldo Jabor, who defended the view opposite to his. I couldn't hold back and got into a discussion. I wanted him to hear that he needed to take into account the fact that part of what nowadays is not only the racial consciousness movement but also the specific hip-hop movement, to which he is affiliated, suffers from a Brazilian desire anxious to imitate the Americans. And in a certain way that fact reaffirms a humiliation of Brazilians in face of the Americans which is no different from the humiliation of Blacks in face of the whites.
> (2006: p. 13)

11. For a history of *négritude* in *paulista* rap, see Pardue (2004).

12. There are many similarities between Ferréz's *Capão Pecado* and Paulo Lins's *Cidade de Deus* (1997). Ferréz recognizes that Lins influenced his work. At the same time he also compares his writing condition as an 'insider' to Lins's university formation (Lins was from Cidade de Deus but had left, and by the time he came to write *Cidade de Deus* he was a postgraduate student doing research in anthropology).

13. Brazilian soccer is also described in the national press and popular culture as being more 'creative' than its European counterpart. The whole subject has been taken up by the São Paulo composer and academic José Miguel Wisnik (who has worked closely with Caetano Veloso), in a forthcoming study which calls into question received interpretative models.

14. For a good analysis of this aspect of US rap, see Davarian Baldwin (2004).

15. See Charisse Jones (2005) for a discussion of this subject in the US context.

16. Ciudad Juárez, in Mexico, is a remarkable counter-example, as the mysterious death of hundreds of young women has been connected, according to at least some interpretations, to drug trafficking. For an enlightenment on this horrendous matter, see Lourdes Portillo's documentary *Señorita extraviada*.

17. See Jean Franco (2002: 222–33) for an excellent discussion of the relations between consumerism and a crisis in masculinity in Latin American cities.

18. See Alba Zaluar (2001) for a description of the *mafiosi* hierarchy and rules inherent to drug traffic.

19. Most discussions of rap gender issues in the US point to female rap as an antidote to male rap's sexism. In fact, female rappers like Queen Latifah, Salt 'N' Pepa, and MC Lyte often play with men's insecurities on the sexual front, and on the consumer front as well. In Brazil, the female rap scene is growing immensely, but the performers have yet to become commercially successful, like their male counterparts.

6 WRITING ON THE WALL AND OTHER INTERVENTIONS

1. The film is based on a novel by Paulo Lins.

2. Neither the women nor their houses were identified by Rossell. The exhibition, however, generated intense debates in Mexico, and several newspaper articles took pride in revealing the identities of the women. See Gallo 2004: 47–69.

References

Aguiar, Flávio. 'O mal-estar da linguagem', *Zero Hora*, 2 March 2002: 7.

Amin, Ash. 'The economic base of contemporary cities', in Gary Bridge and Sophie Watson (eds) *A Companion to the City*. Oxford: Blackwell, 2000.

Andrade, Elaine Nunes de. *Movimento Negro Juvenil: um estudo de caso sobre jovens rappers de São Bernardo do Campo*. Master's thesis, Faculdade de Educação, Universidade de São Paulo, 1996.

Andrade, Mário de. *Macunaíma*. Translated by E. A. Goodland. New York: Random House, 1984.

—— 'Pauliçeia Desvairada'. *Poesias Completas*. Edição Crítica de Diléa Zanotto Manfio. Belo Horizonte; Rio de Janeiro: Villa Rica, 1993.

Aréchiga Córdoba, Ernesto. *Tepito: del antiguo barrio de indios al arrabal*. Mexico: UNIOS, 2003.

Arguedas, José María. *El zorro de arriba y el zorro de abajo*. Buenos Aires: Losada, 1971.

Aridjis, Homero. *La leyenda de los soles*. Mexico: Fondo de Cultura Económica, 1993.

Azuela, Mariano. *La malhora*. Mexico: Botas, 1941.

Báez Rodríguez, Francisco. '40 aniversario de Canal 11', *Etcéter@. Una ventana al mundo de los medios*. http://www.etcetera.com.mx/tv34.asp

Baldwin, Davarian. ' "Black Empires, White Desires": The Spatial Politics of Identity in the Age of Hip-Hop', in Murray Forman and Mark Anthony Leal (eds) *That's the Joint: The Hip-Hop Studies Reader*. London: Routledge, 2004.

Barajas, Rafael (El Fisgón). *Cómo sobrevivir al Neoliberalismo sin dejar de ser mexicano*. Mexico: Grijalbo, 1996.

Baudelaire, Charles. *Art in Paris, 1845–62*. Edited and translated by Jonathan Mayne. New York: Phaidon, 1965.

—— 'The Painter of Modern Life', in *The Painter of Modern Life and Other Essays*. Trans. Jonathan Mayne. New York: Da Capo Press, 1986.

—— *The Flowers of Evil and Paris Spleen*. Poems by Charles Baudelaire. Trans. William H. Crosby. Rochester, NY: BOA, 1991.

Benítez, Fernando. 'Aquí nos tocó vivir', in *Escritoras de Hispanoamérica*. http://redescolar.ilce.edu.mx/redescolar/memorias/escritoras_hispano01/clcristinap.htm

Benjamin, Walter. *The Arcades Project*. Trans. Howard Eiland and Kevin McLaughlin. Cambridge, MA: Belknap, 1999.

—— 'The Return of the *Flâneur*', in *Selected Writings*, vol. 2. 1927–1934. Cambridge, MA: Belknap, 1999, pp. 262–7.

Bennett, Andy. 'Hip-hop am Main, Rapping on the Tyne: Hip-hop Culture as a Local Construct in Two European Cities', in *Popular Music and Youth Culture: Music Identity and Place*. Basingstoke: Macmillan Press, New York: St. Martin's Press, 2000, pp. 133–65.

Berman, Marshall. *All That Is Solid Melts into Air: The Experience of Modernity*. New York: Penguin, 1988.

Blanco, José Joaquín. *Álbum de pesadillas mexicanas. Crónicas reales e imaginarias*. Mexico: Era, 2002.

Bourdieu, Pierre. 'La "jeunesse" n'est qu'un mot', in *Questions de Sociologie*. Paris: Minuit, 1984, pp. 143–54.

Bregantini, Daisy. 'Caetano Veloso é Verbo e Adjetivo', *Cult* 105(9) (August 2006): 11–17.

Bridge, Gary and Watson, Sophie. 'City Economies', in Gary Bridge and Sophie Watson (eds) *A Companion to the City*. Oxford: Blackwell, 2003.

Brito, Fausto. 'O Deslocamento da População Brasileira para as Metrópoles', *Estudos Avançados* 20(57) (2006): 221–36.

Brotherston, Gordon. *Book of the Fourth World: Reading the Native Americas through their Literature*. Cambridge: Cambridge University Press, 1992.

—— *Painted Books from Mexico*. London: British Museum Press, 1997.

—— *Feather Crown*. London: British Museum Press, 2006.

Buck-Morss, Susan. *The Dialectics of Seeing: Walter Benjamin and the Arcades Project*. Cambridge, MA: MIT Press, 1989.

Caldeira, Teresa Pires do Rio. *Cidade de Muros. Crime, Segregação e Cidadania em São Paulo*. São Paulo: Edusp, 2000.

Campos, Cândido Malta. *Os Rumos da Cidade. Urbanismo e Modernização em São Paulo*. São Paulo: Editora Senac, 2000.

Castells, Manuel. *The Informational City: Information, Technology, Economic Restructuring and the Urban-Regional Process*. Oxford: Blackwell, 1989.

—— *The Rise of the Network Society*. Oxford: Blackwell, 1996.

Clendinnen, Inga. *Aztecs: An Interpretation*. Cambridge: Cambridge University Press, 1991.

Cortés, Hernán. 'Segunda Carta de Relación' (30 de octubre de 1520), *Cartas y Documentos* Mexico: Porrua, 1963: 72–3.

Costa, Roaleno Ribeiro Amancio. *O Graffiti no Contexto Histórico-Social, como Obra Aberta e uma Manifestação da Comunicação Urbana*. Vol. II. Master's thesis. Departamento de Artes Plásticas da Escola de Comunicações e Artes da Universidade de São Paulo, 1994.

Cross, John. *Informal Politics. Street Vendors and the State in Mexico City*. Stanford, CA: Stanford University Press, 1998.

—— 'Introduction', *International Journal of Sociology and Social Policy*. Special issue on Street Vending in the Modern World, 21(3/4) (2000). http://openair.org/pub/IJSSP/crossintro.htm

Damiani, Ana Luisa. 'Urbanização Crítica e Situação Geográfica a Partir da Metrópole de São Paulo', in Ana Fani Alessandri Carlos and Ariovaldo Umbelino de Oliveira (eds) *Geografias de São Paulo 1. Representação e Crise da Metrópole*. São Paulo: Contexto, 2004, pp. 19–58.

Davis, Diane E. *Urban Leviathan: Mexico City in the 20th Century*. Philadelphia, PA: Temple University Press, 1994.

Davis, Mike. *City of Quartz: Excavating the Future in Los Angeles.* New York: Vintage, 1992.
—— *Planet of Slums.* London: Verso, 2006.
de Certeau, Michel. *The Practice of Everyday Life.* Trans. Steven Rendall. Berkeley, CA: University of California Press, 1984.
Deleuze, Gilles. *Difference and Repetition.* Trans. Paul Patton. New York: Columbia University Press, 1994.
Donald, James. *Imagining the Modern City.* Minneapolis: University of Minnesota Press, 1999.
Engels, Friedrich. *The Condition of the Working Class in England in 1844.* Oxford: Oxford University Press, 1993.
Eseverri, Máximo. 'Amores perros', *Cineismo.* http://www.cineismo.com/criticas/amores-perros.htm
Espinosa López, Enrique. *Ciudad de México. Compendio cronológico de su desarrollo urbano. 1521–1980.* Mexico, author's edition, 1991.
Fenianos, Eduardo Emílio. *São Paulo. Uma Aventura Radical. Expedições Urbenauta.* São Paulo: Univer Cidade, 2002.
Ferréz. *Capão Pecado.* São Paulo: Labortexto, 2000.
—— *Manual Prático do Ódio.* Rio: Objetiva, 2003.
—— 'Antropo(hip-hop)logia'. *Folha de São Paulo.* 05/04/2006. http://www1/folha.uol.com.br/fsp/opinaia/fz0504200609
Forman, Murray. ' "Represent": Race, Space, and Place in Rap Music', in Murray Forman and Mark Anthony Leal (eds) *That's the Joint: The Hip-Hop Studies Reader.* London: Routledge, 2004.
Franco, Jean. *The Decline and Fall of the Lettered City: Latin America in the Cold War.* Cambridge, MA: Harvard University Press, 2002.
Fuentes, Carlos. *Cristóbal Nonato.* Mexico: Fondo de Cultura Económica, 1987.
Gallo, Rubén. *New Tendencies in Mexican Art. The 1990s.* New York: Palgrave Macmillan, 2004.
Gallo, Rubén. Untitled comments. Daniela Rossell 'Las Lomas II', in Rubén Gallo (ed.) *The Mexico City Reader.* Madison, WI: University of Wisconsin Press, 2004.
Galvão, Patrícia. *Industrial Park: A Proletarian Novel.* Trans. Elizabeth and K. David Jackson. Lincoln, NB: University of Nebraska Press, 1993.
García Canclini, Nestor. *Hybrid Cultures: Strategies for Entering and Leaving Modernity.* Trans. Christopher L. Chiappari and Silvia L. López. Minneapolis: University of Minnesota Press, 1995.
—— 'Las cuatro ciudades de México', in Nestor García Canclini (ed.) *Cultura y comunicación en la ciudad de México.* 2 vols. Mexico: Grijalbo, 1998, pp. 19–39.
—— *Consumers and Citizens: Globalization and Multicultural Conflicts.* Trans. and with an Introduction by George Yúdice. Minneapolis: University of Minnesota Press, 2001.
García Michel, Hugo. *Matar por Angela. Novela pasional de crímenes, sangre y acciones desbordadas.* México, DF: Sansores y Aljure, 1997.
Giannetti, Cecília. 'Morando Dentro do Tema', *Jornal do Brasil Online.* 20/04/2003. http://jbonline.terra.com.br/papel/cadernob/2004/04/20/jorcab20040420001.html
Gilroy, Paul. *The Black Atlantic: Modernity and Double Consciousness.* London: Verso, 1993.

Gleber, Anke. 'Female Flanerie and the Symphony of the City', in Katharina von Ankum (ed.) *Women in the Metropolis: Gender and Modernity in the Weimar Culture.* Berkeley, CA: University of California Press, 1994.

Gottmann, Jean. *Megalopolis: The Urbanized Northeastern Seaboard of the United States.* New York, Twentieth Century Fund, 1961.

Guasco, Pedro Paulo Marques. *Num país chamado periferia: identidade e representação da realidade entre os rappers de São Paulo.* Master's thesis. Departamento de Antropologia Social. Universidade de São Paulo, 2000.

Guzik, Alberto. *O que é ser rio, e correr?* São Paulo: Iluminuras, 2002.

—— *Risco de Vida.* São Paulo: Globo, 1995.

Harrison, Marguerite Itamar. 'São Paulo Lightning: Flashes of a City in Luiz Ruffato's *Eles eram muitos cavalos', Luso-Brazilian Review* 42(2) (2005): 150–64.

Harvey, David. *Spaces of Capital: Towards a Critical Geography.* London: Routledge, 2001.

Highmore, Ben. *Cityscapes: Cultural Readings in the Material and Symbolic City.* New York: Palgrave, 2005.

Hoggart, Richard. *The Uses of Literacy: Changing Patterns in English Mass Culture.* Boston: Beacon Press, 1957.

Holston, James. 'Spaces of Insurgent Citizenship', in James Holston (ed.) *Cities and Citizenship.* Durham, NC: Duke University Press, 1999.
http://oncetv-ipn.net/cristina_pacheco/cristina/index.htm
http://www2.uol.com.br/urbenauta/livrosp.htm

Huq, Rupa. *Beyond Subculture: Pop, Youth and Identity in a Postcolonial World.* London: Routledge, 2006.

Jameson, Frederic. *Postmodernism or the Cultural Logic of Late Capitalism.* London: Verso, 1991.

Jesus, Carolina Maria de. *Child of the Dark: The Diary of Carolina Maria de Jesus.* Trans. David St. Clair. New York: Penguin, 1993.

Jones, Charisse. 'Still Hangin' in the "Hood": Rappers Who Stay Say their Strength is from the Streets', *The New York Times*, 24 September 2005, pp. 54–67.

Kantaris, Geoffrey. 'Street Vision in Latin American Cinema'. Stephen Hart and Richard Young (eds). *Contemporary Latin American Cultural Studies.* London: Arnold, 2003. 177–189.

Kasinitz, Philip (ed.) *Metropolis: Center and Symbol of our Times.* New York: New York University Press, 1995.

Kelley, Robin D. G. *Race Rebels: Culture, Politics, and the Black Working Class.* New York: Free Press, 1996.

Kubrusly, Maria Emília. *O Bar de Flora Paixão.* São Paulo: Sette Letras, 2000.

Lefebvre, Henri. *The Production of Space.* Trans. Donald Nicholson Smith. Oxford: Blackwell, 1991.

—— *Rhythmanalysis: Space, Time, and Everyday Life.* London: Continuum, 2004.

Leite, Rogério Proença. *Contra-usos da Cidade. Lugares e Espaço Público na Experiência Urbana Contemporânea.* Campinas: Editora Unicamp, 2004.

León-Portilla, Miguel. *The Broken Spears: The Aztec Account of the Conquest of Mexico.* Boston: Beacon Press, 1992.

Lewis, Oscar. *The Children of Sanchez; Autobiography of a Mexican Family.* New York: Random House, 1961.

Lins, Paulo. *Cidade de Deus.* São Paulo: Cia das Letras, 1997.

Lira, Andrés. *Comunidades indígenas frente a la ciudad de méxico: Tenochtitlán y Tlatelolco, sus pueblos y sus barrios. 1812–1819.* Mexico: Colmex: 1995.

Lonewolf. 'Barrio Bravo Tepito'. *Brown Kingdom.* http://13radicalriders14.blogspot.com/2005/11/barrio-bravo-tepito.html

Machado, Antonio de Alcântara. *Novelas Paulistanas.* Rio de Janeiro: José Olympio, 1973.

Maricato, Ermínia. *Metrópole na Periferia do Capitalismo: Ilegalidade, Desigualdade e Violência.* São Paulo: Hucitec, 1996.

Marx, Karl. *Capital.* Trans. Eden and Cedar Paul. Vol. 1. London: Everyman's Library, 1957.

Mier y Terán Rocha, Lucía. *La primera traza de la ciudad de México 1524–1535.* Vol. 1. México: UAM; Fondo de Cultura Económica, 2005.

Mitchel, Tony. (ed.). *Global Noise: Rap and Hip Hop Outside the USA.* Middletown, CT: Westleyan University Press, 2001.

Mitra, Ananda. 'Voices of the Marginalized on the Internet: Examples from a Website for Women of South Asia', *Journal of Communication.* 54(3) (Sep. 12, 2004): 492–510.

Mollenkopf, John H. and Castells, Manuel (eds) *Dual City: Restructuring New York.* New York: Russel Sage, 1991.

Monsiváis, Carlos. 'Tepito como leyenda'. *Dias de Guardar.* Mexico: Era, 1970.

—— *Los rituales del caos.* México: Era, 1995.

Muñoz, Boris. 'La ciudad de México en la imaginación apocalíptica', in Boris Muñoz and Silvia Spitta (eds) *Más allá de la ciudad letrada: crónicas y espacios urbanos.* Pittsburgh, PA: Biblioteca de América, 2003, pp. 75–98.

Neal, Mark Anthony. 'I'll Be Nina Simone Defecating on Your Microphone: Hip-Hop and Gender', Murray Forman and Mark Anthony Neal (eds.). *That's the Joint. The Hip-Hop Studies Reader.* New York: Routledge, 2004.

Nivón, Eduardo. 'Introducción', in Ana Rosas Montecón and Guadalupe Reyes Dominguez, *Los usos de la identidad barrial.* México: UAM, 1993.

Ochoa, Ana María. *Músicas locales en tiempos de globalización.* Bogotá: Norma, 2003.

O'Gorman, Edmundo. 'Reflexiones sobre la distribución urbana colonial de la ciudad de México', *XVI Congreso Internacional de Planificación y de la Habitación en México (1937).* México: Porrúa, 1975.

Osumare, Halifu. 'Beat Streets in the Global Hood: Connective Marginalities of the Hip Hop Globe', *Journal of American & Comparative Cultures* (JACultC) 2001 Spring–Summer; 24(1–2): 171–81.

Pardue, Derek. 'Putting *Mano* to Music: The Mediation of Race in Brazilian Rap', *Ethnomusicology Forum* 13(2). November 2004: 253–86.

Peixoto, Nelson Brissac (ed.) *Intervenções Urbanas.Arte/Cidade.* São Paulo: Senac, 2002.

Perry, Imani. *Prophets of the Hood: Politics and Poetics in Hip Hop.* Durham, NC: Duke University Press, 2004.

Piccato, Pablo. 'Communities and Crime in Mexico City', *Delaware Review of Latin American Studies* 6(1). June 2005. http://www.udel.edu/LAS/Vol6-1Piccato.html

Pimentel, Spensy. 'Entrevista com Mano Brown'. *Teoria e Debate* 46 (Nov.–Jan. 2001). http://www.fpabramo.org/td/td46/td46_culture.htm

Poniatowska, Elena. 'Ángeles de la ciudad', in *Fuerte es el silencio.* Mexico: Era, 1997, pp. 13–33.

Racionais MCs. *Holocausto Urbano*, RDS Fonográfica RDL 4006, 1990.
—— *Escolha seu Caminho*, RDS Fonográfica, 1992.
—— *Raio X do Brasil*, RDS Fonográfica, 1993.
—— *Sobrevivendo no Inferno.* Cosa Nostra 001, 1997.
—— *Nada como um dia depois de outro dia.* Zâmbia 050, 2002.
Ramírez, Armando. *Violación en Polanco.* Mexico: Grijalbo, 1979.
—— *Tepito.* Mexico: Grijalbo, 1983.
—— *Chin Chin el teporocho.* Mexico: Grijalbo, 1985.
—— *Bye Bye Tenochtitlan.* Mexico, Grijalbo, 1991.
—— *Me llaman la Chata Aguayo.* Mexico: Grijalbo, 1994.
—— *Sóstenes San Jasmeo.* Mexico: Grijalbo, 1997
—— *La casa de los ajolotes.* Mexico: Oceano, 2000
—— *¡Pantaletas!* Mexico: Oceano, 2001.
Ramos, Julio. *Divergent Modernities: Culture and Politics in Nineteenth-century Latin America.* Trans. John D. Blanco. Durham, NC: Duke University Press, 2001.
Reguillo Cruz, Rossana. *Emergencia de culturas juveniles. Estratégias de desencanto.* Bogotá: Norma, 2000.
Resende, Beatriz. 'São Paulo, SP'. *Babel Livros* 3. October 2001. http://babel.no.com.br
Rheda, Regina. *Arca sem Noé: Histórias do Edifício Copan.* Rio: Booklink, 2002.
Rico, Maite. 'Tepito barrio bravo de Mexico'. *El país* 21 June 2006. http://www.eco.utexas.edu/~archive/chiapas95/2006.06/msg00311.html
Rio, Eduardo del (RIUS). *Quetzalcoatl no era del PRI.* Mexico: Grijalbo, 1987.
—— *500 años fregados pero cristianos.* Mexico: Grijalbo 1992.
Rojas, Carlos. 'Tepito, nariz de la capital'. www.literaturainba.com/escritores/armando_ramirez.htm
Rolnik, Rachel. *A Cidade e a Lei. Legislação, Política Urbana e Território na Cidade de São Paulo.* São Paulo: FAPESP / Studio Nobel, 1997.
Rosales Ayala, Héctor. *Tepito, ¿barrio vivo?* Mexico: UNAM, 1991.
Rose, Tricia. *Black Noise: Rap Music and Black Culture in Contemporary America.* Middletown: Wesleyan University Press, 1994.
Rossell, Daniela. *Ricas y famosas.* Madrid: Turner, 2002.
Ruffato, Luiz. *Eles Eram Muitos Cavalos.* São Paulo: Boitempo, 2001.
Sánchez, Raymundo. 'Veneran en Tepito a Santa Muerte con procesión'. *La crónica de hoy* 25 March 2005. http://www.cronica.com.mx/nota/php?id_nota=173309
San Filippo, Maria. 'Amores Perros'. *Senses of Cinema.* http://sensesofcinema.com/contents/01/13/amroes.html
Sassen, Saskia. *The Global City.* Princeton, NJ: Princeton University Press, 2001.
Savlov, Marc. 'Amores Perros', *The Austin Chronicle.* http://www.austinchronicle.com/gbase/Calendar/Film?Film=oid%3a141094
Schell, Patience A. *Church and State Education in Revolutionary Mexico City.* Tucson: University of Arizona Press, 2003.
Silva, Janice Theodoro da. *São Paulo 1554–1880. Discurso Ideológico e Organização Espacial.* São Paulo: Moderna, 1884.
Smith, Paul Julian. *Amores perros.* London: BFI, 2003.
Soja, Edward. *Postmetropolis: Critical Studies of Cities and Regions.* Oxford: Blackwell, 2000.
Somekh, Nadia. *A Cidade Vertical e o Urbanismo Modernizador.* São Paulo: Edusp; Studio Nobel; FAPESP, 1997.

Sutcliffe, Anthony. 'The Giant City as a Historical Phenomenon', in Theo Barker and Anthony Sutcliffe (eds) *Megalopolis: The Giant City in History*. Houndmills and London: Macmillan Press, 1993.

Tester, Keith. *The Flâneur*. London: Routledge, 1994.

Thrift, Nigel. 'Driving in the City', *Theory, Culture & Society* 21 (2004): 41–59.

Valle, Perla. *Ordenanza del Señor Cuauhtemoc*. Mexico: Gobierno del DF, 2000.

Vice, Jeff. 'Amores Perros'. http://deseretnews.com/movies/view/1,1257,195000072,00. htmls

Villoro, Juan. *El disparo de Argón*. Madrid: Alfaguara, 1991.

Villoro, Juan. *Materia dispuesta*. Mexico, DF: Alfaguara, 1996.

Virilio, Paul. 'The Last Vehicle', in *Looking Back at the End of the World*. Trans. David Antal. New York: Semiotext(e), 1989.

—— 'The Overexposed City', in *The Lost Dimension*. Trans. Daniel Moshenberg. New York: Semiotext(e), 1991.

Vollmer, Günther. *Geschichte der Azteken*. Der Codex Aubin. Berlin: Mann, 1981.

Williams, Raymond. *The Country and the City*. London: Chatto and Windus, 1973.

Wilson, Elizabeth. *The Sphinx in the City: Urban Life, the Control of Disorder, and Women*. Berkeley, CA: University of California Press, 1992.

Zaluar, Alba. 'Violence à Rio de Janeiro: styles de loisirs, de consommation et the trafic de la drogue', *Revue internationale des sciences sociales* 169 (Sept. 2001): 407–17.

Zantwijk, Rudolf van. *The Aztec Arrangement*. Norman, OK: University of Oklahoma Press, 1985.

Zeni, Bruno. *O Fluxo Silencioso das Máquinas*. São Paulo: Ateliê, 2002.

Zola, Emile. *L'assommoir*. Trans. Margaret Mauldon. Oxford: Oxford University Press, 1995.

Index

16060 72, 145
Abramovich, Fanny 24, 27
accident 42–50, 55
acequias 86
Adorno, Theodor 22
Africa 5, 46–7, 129
airport 15, 30, 120
alcoholism 28, 94, 118, 121, 123, 127, 142
Alvarado, Marques de 93
Amores perros 9, 23, 40–56, 59
Andrade, Eliane Nunes de 123
Andrade, Mário de 13–14, 71, 143–4; *Macunaíma* 14, 144; *Paulicéia Desvairada* 13, 71
Andrade, Oswald de 14, 71
Antonio, João 143
apartheid 129; *see also* race
Aquí nos tocó vivir 9, 24, 67–9
archaeology 2, 11, 16–17, 91, 97, 105, 146
architecture 2, 10, 41, 53–4, 74, 100, 150
Aréchiga Córdoba, Ernesto 80, 86–7, 89
Arguedas, José María 96
Aridjis, Homero 146
Aristotle 23–5
Asia 5, 46
Athaíde, Celso 135
automated tellers 35
avant-garde 13–14, 75, 105
Axayacatl, Emperor 85
axolotl 91
azotea 54

Aztec 2, 9–11, 15–16, 85–6, 91, 93, 97, 101, 105, 146, 152–3; empire 10, 85, 93
Azuela, Mariano 88

Babel 46–7
bairro see neighbourhood
Balzac, Honoré de 71
bandeirantes 12
Barajas, Rafael 152
Barbosa, Adoniran 143
barrio see neighbourhood
bars 29, 121, 123, 131, 146
Baudelaire, Charles 9, 21–2, 38, 56, 58, 60–1, 64, 73, 75
beggars 28, 61, 104
Belloto, Tony 145
Benítez, Fernando 68
Benjamin, Walter 7–8, 22–3, 57–8, 64, 67, 71, 145, 147
Benjor, Jorge 118
Berlin 23, 25, 44, 57
Berman, Marshall 75
Blanco, José Joaquín 103–5
Boston 3
boundaries 3, 10, 80, 92, 115
Bourbons 87, 104
Bourdieu, Pierre 110
boxing 64–5, 89, 106, 134
break dancing 110–11, 143
Bridge, Gary 157n4
Brito, Fausto 15
Brotherston, Gordon 91, 157n7, 157n10
Buck-Morss, Susan 22

building legislation 13–14, 74, 108
bus 12, 14, 34, 42, 44, 46, 62–3, 93, 117, 126–7, 141, 149, 152, 159

Caillois, Roger 71
Cairo 81
Caldeira, Teresa Pires 4, 7, 13–14, 54, 74, 114, 141, 154
Callejón de los Milagros 41, 81, 93
Campos, Cândido Malta 2, 12–14
Canal Once (television station, Mexico) 67–9
canals 10, 15–16, 93
cangaceiro 96, 110
Capão Redondo 5, 9, 73–85, 107–10, 112–15, 118, 122–7, 131, 133–5, 140, 142–4; Favela do Fundão 113, 131; *see also* neighbourhood
car 14–15, 44, 49–53, 55–6, 61, 63, 72, 74, 96, 115, 117, 120, 136; car chase 41–4, 54; flows 1, 45, 49–50, 53, 59, 61; noise of 41, 49–50, 112; pollution of 33; and fragmentation 22, 33, 51; and city space 31, 36, 41–4, 50–3, 59–60, 63, 112, 149
Cárdenas, Lázaro 11
Cardoso, Fernando Henrique 157n2
Cascão (rapper) 125
Castells, Manuel 4, 50, 82
Chambers, Ian 21
chaos 30, 66, 88
Chaplin, Charles 25
Chiapas 17
chinampa (water agriculture) 11
Christianity 12, 66, 86, 91, 93, 152
Cia. City 13
Cidade de Deus 145
citizenship 6, 60, 63, 101–2
class 50, 53–4, 57–9, 67–9, 72–3, 84, 101, 110, 114–19, 123, 125, 136–7, 146, 154; division 11, 14, 28–9, 45–8, 52, 67, 73, 75, 82, 88, 93, 95, 116–18; privilege 47–8, 67, 73, 75, 93, 117; *see also* poverty
Clendinnen, Inga 60
cocaine 118, 132
codices 10, 91, 152–3
coffee: elites 12–13

Coixtlahuaca 10, 85
Columbus, Christopher 10, 71–2, 146, 152
Compton 113
computers 35–6, 50
Conceito Moral (rapper) 109
conquistador 43
consumerism 7, 22, 84, 100–2, 110, 117, 120, 136, 138–9, 142
Cooper, James Fenimore 71
corruption 31, 40, 81, 93, 99–100, 146; police 42; politicians and 64, 95, 99, 106
Cortés, Hernán 2, 10, 16, 43, 55, 85–6, 93
Costa, Roaleno Ribeiro Amâncio 149
creativity 90, 103, 124, 130, 132–4, 142, 149
Cross, John 89, 95–6, 99–100
Cuauhtemoc 17, 85–6, 93
cultural life 8, 82, 146
culture of poverty 2, 79, 81, 84, 100, 109–10, 111–12
Curitiba 72

dance 37–8, 62–3, 109–11, 116, 127, 129–30, 143, 146, 150–2; *see also* break dancing
Dantas, Audálio 109
Darwin, Charles 71
Davis, Mike 5, 54, 121, 154
Davis, Diane 11
de Certeau, Michel 7, 51–2
de Chirico 9
death 28, 30, 38, 43, 60–3, 72, 92–3, 107, 119–20, 122–4, 126, 128, 130, 132, 138–42
Debord, Guy 66
Del Rio, Eduardo 152
Deleuze, Gilles 6, 25, 29–31
de-industrialization 5, 147–8; *see also* unemployment
detective novel 71–2, 145–6
Díaz, Bernal 11
Díaz, Porfírio 11
Dickens, Charles 2, 95, 97–100, 102
dikes 10
DJs 41, 49, 111, 113

dog 28, 41–7, 55, 61; 'aperreamiento' 43; fighting 42–3, 47–8, 50, 52–4; and security 28, 53; as displaced violence 43, 47, 52
drug 28, 62, 68, 85, 115–16, 122–3, 125, 128, 130, 134, 138; traffic 53, 74, 90, 107, 119–21, 135, 139–42; economy 53, 119
dystopia 146

education 7, 53, 56, 58, 67–8, 70, 75, 83, 91, 99, 103, 109, 113, 119, 126, 130, 137, 140, 143
Engels, Friedrich 6
Espinosa López, Enrique 10–11
Europe 10, 41, 86, 109; and the invasion of America 2, 12, 15–17, 23, 41, 55, 84, 86–8, 92, 107, 116

factory workers 14, 25, 59, 108, 123
farce 145
favela 74, 107, 109–10, 113–14, 117, 119–20, 122–3, 126, 128, 131, 134, 136, 141
fayuca 84, 90, 102
Fenianos, Eduardo Emílio 9, 24, 69–75, 143
Ferréz 9, 109–10, 114, 122–6, 130, 132–5, 138–40, 142–3; *Capão Pecado* 109, 122–5, 129–30, 135, 138, 140, 142; *Manual Prático do Ódio* 109, 139
films: *16060* 72, 145; *Amores perros* 9, 23, 40–56, 59; *Babel* 46–7; *Callejón de los Milagros* 41, 81, 93; *Cidade de Deus* 145; *Modern Times* 25; *Nosotros los pobres* 81; *O Homem que Virou Suco* 5; *O Invasor* 145; *O Príncipe* 23; *Opressão* 24, 62–4, 75; *Pulp Fiction* 41; *Run Lola Run* 44; *São Paulo: Sinfonia da Metrópole* 14; *Sinfonie einer Großstadt* 14, 25; *Sliding Doors* 44
flâneur 25, 29, 38, 66–7; in the modern city 6–9, 57–8, 145; and gender 62–4; and citizenship 22–3, 60, 64; and class difference 56–7, 61–2, 67, 75, 117, 119; television zapping as 9, 66–7, 69, 72–3, 75

floods 10
Folha de São Paulo 107–8, 110
Forman, Murray 111, 113, 137
fragmentation 3, 6–9, 21–3, 25–7, 30–4, 38–9, 44–9, 51–3, 55, 84, 136, 145
Franciscans 86
Franco, Jean 5, 64
Fuentes, Carlos 21, 146

Gallo, Rubén 154, 156
Galvão, Patrícia 14, 33, 108, 143
Gamio, Manuel 11
gang wars 114–15, 119, 149
García Canclini, Nestor 7, 17, 21, 23, 82, 93, 101, 136
García Bravo, Alonso 10
García Michel, Hugo 146
Gaspar, Frei 12
gated communities 114, 154, 156
gaze 24, 56, 61, 91, 117
gender 14, 62–4, 75, 95, 98, 137–9, 144, 146
generality 29–30
geography 3–4, 7, 9, 12, 57, 84, 91–3, 100, 114, 134–6, 140–3, 145
ghetto 111–13, 133, 140
Gilroy, Paul 112, 129
Gleber, Anke 63
globalization 4–6, 21, 26, 36, 41, 46–7, 49, 52, 82–3, 94, 105, 110–12, 119–21, 133, 142; and culture 110–11, 119, 121; and economy 4, 50, 105, 119, 133
Gottmann, Jean 3
Gramsci, Antonio 136–7
graffiti 17, 37–8, 109–11, 113, 115, 148–51
Guadalajara 65
Guadalupismo 64, 66
Guasco, Pedro Paulo Marques 114, 121–4, 127
Guatemala 10, 17
guerrilla 42, 46, 56, 60
Guzik, Alberto 146

hand-held camera 41, 45, 54
Harrison, Marguerite 26–7, 31–2
Harvey, David 4, 7

haussmanization 14
Hegel, G. F. 30
helicopters 27, 29
Highmore, Ben 58
hip-hop 109–13, 120, 123, 133, 135,
    137, 139, 142; *see also* rap
Hoggart, Richard 80
Holston, James 6
'hood' *see* neighbourhood
homosexuality 26, 36–7, 48, 62, 64;
    homophobia 137
Humboldt, Alexander von 11
Huq, Rupa 110
hydrography 15

Iannarone, Antonio 151
Ice Blue 109, 111, 134
Ice Cube 111
illegality 53, 74, 84, 89–90, 99, 102,
    119–20, 149
IMF 5, 52
immigration 5, 11, 12, 14, 15, 107, 152;
    European 2, 12, 92; Italian 14, 81,
    84, 108; Indigenous (Mexico City)
    17; from rural regions (São Paulo
    and Mexico City) 5, 11; from
    Northeast of Brazil 5, 15, 84–5, 108
inapprehensibiliy 21, 23
incompletion 26–7, 31–2
Independence: Brazil 70; Mexico 11, 87
Indigenous people 12, 17, 80, 86–8,
    91, 94, 100, 103–5, 107, 116; and
    property 11, 74, 86–7
industrialization 5–6, 11, 13–14, 85
informal economy 5, 53, 89–90, 92, 98,
    103, 105, 121; *see also* unemployment
inner city 84
internet 9, 29, 36–7, 41, 72, 110, 114,
    124–6, 132–4, 139–40, 142–3
invisibility 34, 48, 56, 64, 68

Jameson, Fredric 6
Jesuits 2, 12, 107, 134
Jesus, Carolina Maria de 109, 126,
    143
Jornal do Brasil 110
Joyce, James 6, 21, 30, 97
Juárez, Benito 11
jungle 23, 69–72

Kantaris, Geoffrey 48, 52–3
Kelley, Robin D. G. 112
Kemeny, Adalberto 14
kidnapping 29, 50–2, 92–3
Kierkegaard, Sören 29
Klee, Paul 7
Kraniauskas, John 48
Kubrusly, Maria Emília 146

La onda 5
labyrinth 22, 95
Lady Rap 111, 114
Lajolo, Marisa 158n2.3
language 10, 16–17, 21–2, 31, 34, 36,
    39–41, 65, 67, 80–1, 83, 85, 88, 91,
    94, 108, 112, 115, 133, 143–4, 151–3
lawlessness 89
Lefebvre, Henri 7, 25, 83–4, 97
Leite, Rogério Proença 7, 90
Leñero, Vicente 81
Leon-Portilla, Miguel 86
Lewis, Oscar 81, 84
Lins, Paulo 158n8, 160n1
Lira, Andrés 87
London 2, 44, 59
loneliness 22, 25, 28, 58, 94; *anomie* 55
Lopes, Leonardo 110, 133
López Portillo, José 105
lumpenproletariat 28, 46
Lustig, Rudolf Rex 14

Machado, Antonio de Alcântara 14,
    78, 108, 143
Maia, Francisco Prestes 14
Malcolm X 129
Mallarmé, Stepháne 6, 30
Mano Brown 109, 111, 118–19, 122–4,
    127–37
mapping 6, 21, 69, 80, 115, 119, 124,
    145
marginality 61, 67, 88, 91, 110, 121,
    125, 129, 154
Maricato, Ermínia 14, 74, 107
Marinetti, Filippo 33
markets 52, 84–6, 89, 92, 94–5, 97, 99,
    101–2; pre-Columbian 2; *see also*
    street vending; *see also tiangui*
Marco Polo 72
Marvin Gaye 131

Marx, Karl 7, 22, 25, 28, 84, 101
masses 5, 25, 28–9, 31, 60, 64–5, 89, 152
Maximilian, Emperor 87
Maya 10, 151–2
MCs (Masters of Ceremony) 111, 113, 115, 131
media 7–8, 12, 33, 39, 45, 47–9, 64, 68–9, 71–3, 85, 92, 100–1, 110, 119, 121, 131, 136, 138–9, 142, 153
Meireles, Cecília 24
memory 24, 57–8, 60, 98, 120, 144
Mesoamerica 10, 16
metro 12, 16–17, 34, 64–5
Mexica 9–10, 17, 60, 152
Mexico City 1, 3–10, 15–17, 21, 23–4, 40–52, 55–6, 59, 64–9, 75–84, 86–9, 92–8, 100, 103, 105, 145–7, 149–54; Angel de la Independencia 67; Avenida Guerrero 91; Calle Niño Perdido 58; Centro Histórico and Zócalo 2, 5, 40–1, 57, 59, 64, 84–5, 91–2, 99–103, 105, 153; Cerro de la Estrella 93; Ciudad Nezahualcoyotl 5, 60, 93; Colonia Roma 55, 92; Condesa 55, 92; Coyoacan 10, 17, 92; Eje Vial 59; Gran Canal de Deságüe 15, 93; Iztapalapa 17, 90, 93; La Lagunilla 92; La Merced 91, 104; Museo Nacional de Antropología 2, 16–17; Palacio de Bellas Artes 92; Palacio Nacional 91; Paseo de la Reforma 2, 67, 104–5; Polanco 92–3, 154; Santa María la Redonda 58–9; South of 10, 57, 59, 92–3; Tacubaya 10; Templo Mayor 2, 16–17, 91, 105, 146; Torre Latinoamericana 41, 43, 59, 92; Villa de Guadalupe 59, 68; Xochimilco 11; as Distrito Federal (DF) 2, 11, 15; foundation of 2, 9–10, 15–16; lakes 9–10, 11, 15–16, 55, 86, 93; palaces 10–11; siege of 10, 86; *traza* 10, 86; volcanoes 1, 16, 41, 55; *see also* Tenochtitlan; Tepito; Tlatelolco
Mier y Terán Rocha, Lucía 10, 86
Minas Gerais 13, 24
minimum wage 11, 73
misogyny 137–8

Mitchel, Tony 111
Mitra, Ananda 140
Mixtec 17
mobile phones 65, 70, 112, 131
Moctezuma 85, 93
Modern Times 25
modernist city 6, 13, 21–2, 25, 29–30, 33–4
Molina, Alonso 85
moment 36–7, 43–7, 49, 55, 120, 154
Monsiváis, Carlos 9, 16, 23, 55, 64–7, 75, 89, 106, 134, 152
monster city 6, 23, 30, 39, 66, 93
montage 22, 24
multitude 1, 22, 60, 64–7, 89
Muñoz, Boris 66
mutirão 74, 143
MV Bill 135

NAFTA 17
Naguib, Mahfouz 81
Nahuatl 10, 17, 85, 88, 93, 97, 152–3
Neal, Mark Anthony 137
negritude 115, 129
neighbourhood 4, 6, 28, 40, 54–5, 59, 73, 75–89, 94–7, 101, 107–9, 111–19, 121, 124, 126–40, 143–4, 147, 149–50; as *bairro* 6, 73, 75–81, 83–4, 107, 109–10, 113, 115, 118, 133–4, 138, 140, 144; as *barrio* 6, 9, 79–89, 91–2, 94, 99, 104, 106, 134, 144; as *barrio de índios* 86–7; and identity (identidad barrial) 9, 12, 40–1, 45, 79–80, 82–3, 88, 91–2, 94, 133; wealthy 12–13, 28, 45, 55, 73–4, 92, 114–15, 117, 129, 141, 154, 156; working-class 4–5, 11–12, 14, 28, 40, 45, 108, 147; *see also* Capão Redondo; Tepito; São Paulo: Bixiga
neo-liberalism 52–3, 98, 106
New Fire ceremony 93
New York 79, 112
Nicaragua 17
Nietzsche, Friedrich 29–31, 37
night life 57, 61, 146
Nivón, Eduardo 81
noise 43, 41, 49–50, 61; machine 43; music as 43, 49, 112, 115, 131–2

Nosotros los pobres 81
nostalgia 16, 23, 57, 80
Notorious B. I. G. 112
Novo, Salvador 85
Nueva España 10, 92

O Homem que Virou Suco 5
O Invasor 145
O'Gorman, Edmundo 86
Oaxaca 11, 17, 65
Ochoa, Ana María 102
Olivera Ramos, Jorge 103–4
Olmec 10, 16–17, 151
oppression 63
Opressão 9, 24, 62–3, 75
Osumare, Halifu 111
Otomi 17

Pacheco, Cristina 9, 24, 67–9, 75
Pacheco, José Emilio 1, 16
Paraguay 12, 17, 60
paranoia 56, 116
Paris 2, 14, 21, 23, 73
parks 27, 70, 115, 117–18, 125
Peixoto, Nelson Brissac 147–8
periferia 9, 12, 14–15, 33, 74–5, 85,
    104, 107–9, 114–15, 119–22, 124,
    127–8, 130, 132–3, 138, 142–3
Perry, Imani 114, 132
phantasmagory 22, 64, 147
photography 45, 47, 49, 53, 114, 125,
    149, 151, 154–6
Piccato, Pablo 90
piracy 84, 90, 102–3
playboy 115, 118, 121, 123, 125–6,
    129–30, 134
police 28, 42, 46–7, 50, 59–63, 75,
    93, 98, 102–3, 106, 120, 123–4,
    130–1, 142, 148, 150, 156; *see also*
    corruption
pollution 23, 25, 40, 54, 66, 82, 149,
    157; water 15, 70, 72; air 15, 23, 33,
    152
Poniatowska, Elena 5
posse 109
post-industrial economy 4, 6, 113, 121,
    148; *see also* deindustrialization;
    unemployment
postmodernity 4–6

poverty 2, 48–9, 66, 75, 81, 84, 89, 91,
    97, 108, 114, 124, 128–9, 133, 143,
    153–4; *see also* class division
PRI (Partido Revolucionario
    Institucional) 95, 99–100, 103, 106,
    152, 154
proletarian 14, 25, 28, 30, 108, 142
property value 80, 108
prostitution 28, 57, 61, 63, 68, 81, 104,
    138
Proust, Marcel 97–8
PT (Worker's Party) 114, 135, 149
Public Enemy 111, 129
Pulp Fiction 41
pulque 86, 94

race 30–1, 79–82, 93, 112, 115, 128–30,
    144; racism 62–3, 74–5, 104–6, 108,
    120, 126, 129, 134
Racionais MCs 109, 111–15, 117–21,
    123–7, 129–34, 136–40, 143
radio 24, 30, 33–4, 39, 52, 62, 71, 112,
    131, 135, 142
railways 59
Ramalho, João 12
Ramírez, Armando 9, 24, 57–8, 64, 67,
    75, 82–5, 87–8, 90–4, 97, 99–103,
    105, 146; *Tepito* 9; *Chin Chin el
    teporocho* 91; *Bye Bye Tenochititlán*
    57; *Me llaman la Chata Aguayo* 91,
    94–105; *Sóstenes San Jasmeo* 91–2,
    146; *La casa de los ajolotes* 91, 146;
    ¡Pantaletas! 102–3, 105
Ramos, Julio 5
rap 9, 32, 73, 82–4, 107, 109–15,
    118–19, 123–8, 140–7; East Coast
    112; gangsta 113, 137; West Coast
    112, 113
recording industry 90, 102, 111
Reguillo Cruz, Rossana 7, 82, 110
repetition 25, 29–32, 42–3, 73, 129,
    139–40, 142
Republic, proclamation of (Brazil)
    13
Revolution (Mexican) 11, 17, 86, 88,
    152
Reyes, Alfonso 11
Reyna, Fernando 88
Rheda, Regina 146

rhythm 1, 14, 21, 25, 28, 30–4, 58, 111, 116, 118, 120, 131, 146
Rio de Janeiro 13, 15, 16, 135
rituals 2, 64, 66
Rivera, Diego 17, 100
Rodríguez, Fernando Báez 69
Rojas, Carlos 83, 90–1, 93
Rolnik, Rachel 13
Rome 23, 37, 86
Roque, Fátima 149–50
Rosa, João Guimarães 96, 110
Rosales Aylada, Héctor 87–8
Rose, Tricia 111, 113, 137
Rossel, Daniela 154–6
Ruffato, Luiz 9, 23–5, 28, 30–3, 36, 38–9, 145
ruins 2, 10, 16–17, 91, 104–5, 147–8
Rulfo, Juan 96
Run Lola Run 44

Sabotage 118
Sahagún, Fray Bernardino de 92
Sainz, Gustavo 5
Salamanca 86
Sales, Milton 111
sampling 111, 118, 133
San Lorenzo 81–2
San Francisco 152
Sánchez, Raymundo 90
Santo André da Borda do Campo 12
Santos 2
São Paulo 1–9, 12–17, 23–5, 27–30, 32–3, 36–40, 51, 54, 62, 64, 69–81, 84–5, 107–15, 117–18, 120, 124–6, 130–1, 133, 141–51, 154; ABC 12, 15; Alto de Pinheiros 73; Anhangabaú 17, 36; Avenida 9 de Julho 33; Avenida Brasil 27–9; Avenida Faria Lima 33; Avenida Paulista 13–14, 33, 37, 74, 149; Avenida Rebouças 27, 29, 33, 149; Bixiga 84, 108; Brás, 108, 147; Brooklin 118; Butantã 126; Cambuci 147, 149–50; Campo Limpo 39, 75; Campos Elíseos 12; Cantareira 70; Guarapiranga 17, 70; Higienópolis 12; Ibirapuera 17, 27; Jardim Ângela 73–5; Jardim Rosana 113; Liberdade 125; Morumbi 114, 118; Parque Ipê 117–18; Pateo do Colégio 2; Pinheiros River 15, 114, 118, 141; Piratininga 12; Riacho do Ipiranga 70; Rua 24 de Maio 27; Santo Amaro 107; Tietê River 15, 17, 33; Zona Sul 5, 39, 70, 73, 107, 111–12, 114–15, 118–19, 122–3, 125, 127–8, 131, 133, 142; foundation of 2, 12, 107; Plano das Avenidas 14
São Paulo: Sinfonia da Metrópole 14
Sassen, Saskia 3–5, 154
schools 2, 12, 24, 28–9, 38, 42, 54, 80, 83, 107, 113, 119, 121, 123, 130, 135
science 71
scratching 111, 115, 131
Semana de Arte Moderna 12–13, 71
servants 48
sewage 34, 120–1, 124
shooting 38–9, 42–4, 56, 62–3, 120, 122
shopping centres 100
Silva, Luiz Ignacio Lula da 135
Silva, Janice Theodoro da 12
Sinclair, Upton 71
Sinfonie einer Großstadt 14, 25
skyscrapers 40, 69
slavery 2, 13, 28, 107, 134; black 110, 112, 121, 129; indian 12
Sliding Doors 44
Smith, Paul Julian 40–1, 43, 46, 48–50, 54
social contrast 6, 67, 117
Soja, Edward 3, 21, 135, 140
Somekh, Nadia 13
Soto, Hernando de 98, 103
space production 6–7, 9, 39, 46–7, 83–4, 97, 112, 114, 122, 132, 135, 147–50
spectacle 65–6, 73
speed 1, 6, 22–3, 29, 33–4, 41, 42
sports 65, 118, 130, 133, 138
street 1, 13, 23, 27–9, 38, 49–50, 52–3, 58–9, 61–4, 74, 89, 94, 115, 118–19, 122–3, 131, 143, 147; lights 25, 34–6, 42, 52, 122; children 2, 47, 60–1, 68; bum 42, 45, 56, 104; vending 29, 68, 84, 86, 89–91, 94–105
Sunstone 17, 146, 152
Superbarrio 80

tagging 115, 149–50
Taibo II, Paco 145–6

taxis 29, 50, 64, 89
telão 33, 39, 45
television 9, 24, 34, 36, 47, 50, 62,
    64–7, 69, 71, 75, 92, 103, 117, 121,
    131, 135, 138, 142, 153
Tenochtitlan 2, 9–10, 15–16, 55, 57,
    83–4, 91, 102, 146, 152–3
Teotihuacan 11, 17, 55, 100
Tepito Arte Acá 91–2, 94
Tepito 9, 11, 17, 57, 59, 79–92, 94–6,
    99, 105–6, 107, 134, 144
Tester, Keith 58
theatre 13, 26–7, 146; sketches 26–9
thieves 28, 61–2, 75, 108, 128
Thrift, Nigel 51
tiangui (tianquiz) 85–6, 94–7, 101, 104;
    see markets (pre-Columbian)
tlacuilolli 10, 16–17, 85, 93, 151–3
Tlatelolco 17, 85–7, 93
toponyms 17, 85, 93, 113, 117, 153
tourism 2, 16–17, 23, 40, 46, 79, 90, 108
tribute lists 85
Tupac Shakur 111–12, 129
Tupi-Guarani 12, 17, 74, 107

unemployment 16, 28, 97, 121, 142
university 12, 64–5, 68, 83, 93, 99, 103,
    130; college 46, 56, 83, 140, 143
unrepresentability 9
urbenauta 9, 24, 69–70, 72, 75, 143

Valle, Perla 86
vecindad 81, 91, 94

Veloso, Caetano 1, 147
video 40, 94
Villas Boas brothers 72
Villoro, Juan 21, 81, 146
Vinci, Laura 147–8
violence 6, 23, 28–9, 39–41, 44, 49,
    52–3, 56–7, 61–2, 79, 89, 93, 107,
    109–10, 115–16, 118–19, 121–3,
    125–6, 128, 130–1, 134–5, 138–9,
    145–7, 154
Virilio, Paul 35–7, 45

walking 7, 21, 24, 37–8, 51, 57–64, 69,
    96; and invisibility 29, 49–50, 55; and
    gender 62–3; and class difference
    60–1, 75; *see also* flânerie
walls 4, 7, 9, 38, 50, 54, 62, 114, 141,
    145, 150–1, 154, 156
Washington DC 3
Watts 113
waves (electromagnetic, radio) 34–5
weapons 53, 112, 119
William, Raymond 21, 80
Wilson, Elisabeth 63
www.capao.com.br 108, 110, 124–6,
    132–4, 139–41, 143

youth culture 82–4, 110

Zeni, Bruno 9, 23, 32–3, 35–7, 39,
    45–6, 146
Zola, Emile 71, 83
Zumbi 110